THE ESSENTIAL SPACE FACTS YOU MUST KNOW

500 Mysteries About the Stars, Moons, Planets, Black Holes, and Beyond

L. S. L'AMAR

The Essential Space Facts You Must Know
500 Mysteries About the Stars, Moons, Planets, Black Holes, and Beyond

Author: L. S. L'amar

ISBN: 9798397289597

Cover design by Irene Martinez
Interior Design by Zakira K.

For more information about this book and other publications, please contact contact@whiteelephants.de

TABLE OF CONTENTS

SECTION 1
Our Cosmic Connection: The Stardust That Makes Us

SECTION 2
The Sun: From its Magnetic Field to Its Explosive Activity

SECTION 3
The Fascinating World of Planetary Formation

SECTION 4
The Incredible Shrinking Planet: Mercury

SECTION 5
Venus: The Scorching Jewel of Our Solar System

SECTION 6
Earth: A Small, Blue Marble in the Vast Universe

SECTION 7
Mars: Our Potential Alternative Home

SECTION 8
Jupiter: The Gigantic World of Gas and Storms

SECTION 9
Saturn: The Ringed Wonder of the Solar System

SECTION 10
Uranus and Its Unconventional Characteristics

SECTION 11
The Blue Neptune and Planet 9

SECTION 12
The Curious World of Dwarf Planets

SECTION 13
The Moon: Exploring Earth's Celestial Companion

SECTION 14
Other Moons in The Solar System

SECTION 15
Space Rocks: Comets and Asteroids

SECTION 16
Time and Space: Journeys, Theories, and Challenges

SECTION 17
The Fascinating Realm of Black Holes

SECTION 18
The Search for Habitable Planets and Signs of Life

SECTION 19
Exoplanets: A Journey Through Eccentricity

SECTION 20
Stars: From Birth to Death

SECTION 21
Galaxies: Exploring Cosmic Marvels and Mysterious Entities

SECTION 22
Cosmic Collisions: Explosive Encounters that Shape the Universe

SECTION 24
Theories, Paradoxes, and Puzzles

SECTION 25
Subatomic World: Quantum Mechanics and Antimatter

SECTION 26
Historical Events: Pioneers, Discoveries, and Future Missions

SECTION 27
Exploring the Future Frontier: Space Tourism and Beyond

SECTION 28
Space Oddities: Eccentric and Curious Journey

INTRODUCTION

The universe, with its eccentricity and infinite possibilities, holds unparalleled significance, shaping our perception of reality and awakening a profound connection to our own selves. It captivates our complex minds, ignites our curiosity, and prompts us to contemplate the very essence of existence. Gazing upon the shimmering stars on an ordinary night, we are confronted with the mind-boggling vastness of the sky, and their sheer magnificence shakes our ordinary perspectives of life, reminding us of our infinitesimal presence in the cosmic order. The countless stars, galaxies, and celestial phenomena not only reveal our size in the adventurous cosmic playground but also remind us of the incredible privilege of being alive in this mysterious reality. This realization humbles us, underscoring that our lives, with all their triumphs and disasters, are part of a much larger narrative unfolding in the cosmic landscape.

The charms of nature stun and sparks our intrinsic curiosity, inviting us to pause and contemplate the very essence of existence. In these moments, we find ourselves pondering the philosophical questions that have intrigued our ancestors for ages. As we explore the cosmic embrace, we strive to understand and find our place within this vast tapestry. Regardless of the path we choose, our destiny is inseparably linked to science. Understanding its concepts is crucial for our existence and future. Those who comprehend are more inclined to enjoy the journey of life.

This book aims to share the most essential facts about the cosmos and space, distilling complex scientific concepts into an accessible and inclusive format that encourages a deeper appreciation and engagement with science. The primary objective of this book is to provide essential and comprehensive wide education about the cosmos in a brief way, from the birth of the universe in the Big Bang to the potential colonization of Mars and much, much more. Within these pages, you will encounter 500 fascinating facts and concepts that not only will stimulate and challenge your intellect but also will push the boundaries of conventional understanding. Let the wonders of science be our guide as we embark on this literal voyage, unraveling the profound beauty and intricacy of the universe.

"

*The universe is a grand
symphony, with each celestial
body playing its unique
note in perfect harmony. To
understand its melody is to
unlock the secrets of existence
itself.*

— Friedrich Nietzsche

SPACE AND UNIVERSE

SECTION 1
Our Cosmic Connection: The Stardust That Makes Us

1. **O**ur **Family Tree Begins in The Death of Stars**. Every atom in our body is made of stardust. In other words, everything on Earth, from rocks, stones, water, and crystals to all living things like people, animals, insects, grass, and flowers, is made of this stardust. Every atom of your DNA, skin, blood, and bones is stardust. Even the oxygen you are breathing right now is made of stardust. It is a fascinating and humbling thought that we are all essentially made up of the same cosmic material that profoundly connects us to the universe and underscores the interconnectedness of all things.

2. **W**here **It All Began: The Big Bang Theory and the Birth of the Universe.** The Big Bang Theory is a widely accepted explanation for the origin of the universe. It proposes that the universe began as an incredibly dense and hot point called a singularity, which suddenly expanded and cooled down around 13.8 billion years ago, eventually giving rise to the diverse and vast universe we see today. The Big Bang Theory has profoundly impacted our understanding of the universe, leading to the development of cosmology. This field seeks to understand the universe's structure, origins, and evolution. It has also opened new areas of study and questions, such as the nature of dark matter and dark energy, which are still not fully understood.

3.   **The Cosmic Microwave Background: Evidence for the Early Universe and the Big Bang Theory.** The cosmic microwave background radiation is the afterglow of the Big Bang, the earliest moment of the universe. It was discovered in 1964 by scientists and astronomers Penzias and Wilson, who were studying radio signals from space. Their discovery was later recognized with the Nobel Prize in Physics in 1978. The cosmic microwave background radiation is exceptionally uniform and isotropic, with a temperature of just about 2.72 Kelvin, making it the coldest thing in the universe. It provided a snapshot of the universe when it was about 380,000 years old before atoms formed and the universe became transparent. The cosmic microwave background radiation is considered one of the strongest pieces of evidence for the Big Bang theory and is still being studied today to learn more about the early universe.

4.   **The Universe was Born.** The Universe as we know it has not existed forever; it was born. Roughly 13.82 billion years ago, which is the age of the universe; all the energy, matter, atoms, and time erupted into an extremely dense point called the Big Bang. Amazingly, we can understand the universe and how it came to be, from the Big Bang to the formation of galaxies, planets, and us.

5.   **The First Moments After the Big Bang.** Right after the Big Bang, something spectacular happened. The universe rapidly expanded and cooled in just a few moments, going from an unimaginable temperature of 10^{32} Kelvin to a "cooler" temperature of around 10^9 Kelvin in less than a second. This translates to approximately 2.5×10^{32} degrees Fahrenheit

to around 1.3 x 10^9 degrees Fahrenheit (1.4 x 10^32 degrees Celsius to around 7.2 x 10^8 degrees Celsius)! As the universe kept expanding and cooling down, protons and neutrons formed. These were followed by the formation of electrons and atoms. This process took around 380,000 years to complete.

6. **The Particle Horizon: The Boundary of the Observable Universe.** The Particle Horizon refers to the maximum distance from which light can have traveled to reach us since the Big Bang. Beyond this limit, light has not had enough time to reach us yet, and therefore, the region lies beyond our observable universe. The Particle Horizon also marks a boundary beyond which the universe's matter and energy density become too high for light to penetrate, effectively making it opaque. This concept was first introduced by astrophysicist George Ellis in 1971.

7. **95% of the Universe is Invisible**. The Overwhelming Presence of the Invisible Universe. It is mind-boggling to realize that 95% of the universe is not visible to us. Dark energy accounts for 68%, followed by dark matter at 27%, and only 5% of normal matter that we can comprehend. Thus, the "full picture" of our reality, including what we are a part of, remains enigmatic to our finite understanding.

8. **The Immensity of the Milky Way.** Next to Andromeda, the Milky Way is the second-largest galaxy in what astronomers call the Local Group. The Milky Way is 105,700 light-years wide, which corresponds to a distance of approximately 1,000,000,000,000,000,000 kilometers. The quickest spacecraft would need 18,000 years to travel one light year, and we're

talking about 100,000 of them with the Milky Way. The vast expanse of the Milky Way truly puts into perspective the size of the little blue dot we call home. It's almost as if we're just a speck of dust in the vastness of space.

9. **The Mind-Boggling Scale of the Cosmos: More Stars Than Grains of Sand.** The cosmos abounds with an inconceivable multitude of stars, transforming it into an endless ocean of shimmering luminosities that stretches beyond our imagination. Our Milky Way galaxy alone holds an estimated 100 billion stars, but that's just the tip of the iceberg. Scientists believe there are a staggering 2 trillion galaxies in the observable universe, each housing billions of stars. To put it into perspective, there are more stars in the universe than grains of sand on all of Earth's beaches.

10. **The Mysterious Side of the Universe: Dark Matter.** The Universe is full of mysteries, and one of the most perplexing is dark matter. When scientists noticed differences in the rotation of galaxies in the 1930s, they hypothesized invisible matter. They discovered that the outer regions moved at the same rate as the inner regions, indicating the presence of matter that telescopes could not detect. Additional research into galaxy clusters and cosmic background radiation has confirmed that dark matter accounts for an astounding 85% of all matter in the universe. Dark matter is believed to be made up of yet-undiscovered particles such as Weakly Interacting Massive Particles (WIMPs) or axions which play a vital role in the formation of large-scale structures such as galaxy clusters and cosmic filaments, as well as in holding galaxies together. Despite extensive research, the true nature of dark matter

remains unknown, and scientists continue to look for clues about its composition and properties.

11. **The Dark Side of The Universe: Dark Energy.** Similarly, dark energy is another mysterious aspect of the Universe that scientists are still trying to understand. Initially detected during the late 1990s, this phenomenon arose when scientists observed that the expansion of the Universe was, in fact, accelerating rather than slowing down. This discovery implied the presence of an invisible force, which scientists dubbed "dark energy." Observations of Type Ia supernovae and the cosmic microwave background have bolstered the case for dark energy, with the European Space Agency's Planck satellite mission providing the most precise measurements to date. However, the true nature of dark energy remains a puzzle, and theories about it range from the cosmological constant to scalar fields. Despite the ongoing research, dark energy continues to challenge our understanding of the Universe and our place within it.

12. **The Mystery Behind the Universe's Accelerating Expansion.** The universe's expansion has been a subject of study since Edwin P. Hubble's discovery in the 1920s, but it wasn't until 1998 that physicists detected a repulsive force known as "dark energy" that was causing the universe to accelerate. This acceleration is still one of cosmology's most perplexing and enigmatic phenomena, and its exact cause remains unknown. New data from the Hubble Space Telescope has led to a surprising discovery - the universe's expansion rate doesn't match our expectations. This inconsistency, known as the "Hubble Tension."

13.   **The Hubble Tension: A Discrepancy in the Universe's Expansion.** The Hubble Space Telescope has revealed a discrepancy in the universe's rate of expansion, known as the "Hubble Tension." The discrepancy appears when contrasting the Hubble constant-measured local rate of expansion with the rate of expansion in the early cosmos as established by measurements of cosmic microwave background radiation. The two measurements are not in agreement, with the local measurement being about 9% higher than the early universe measurement. This disparity could be attributed to previously unknown physics, such as variations in the properties of dark energy or dark matter, or to measurement errors. The Hubble Tension is an enthralling mystery that requires more information.

14.   **The Observable Universe: An Inconceivable Expanse of 93 Billion Light-Years.** The observable universe is a breathtaking entity that astounds us with its sheer enormity, estimated to have a diameter of 93 billion light-years. With billions of stars and galaxies, comprehending their vastness is almost unthinkable. For example, the distance between Earth and the nearest star, Proxima Centauri, is around 4.2 light-years, which is equivalent to about 25 trillion miles (approximately 40 trillion kilometers). While scientists continue to uncover new discoveries about the observable universe, its vastness raises existential questions about our place in the cosmos and the unknown mysteries that lie beyond. The universe is vast, and it can be challenging to comprehend its scale.

15.   **The Cosmic Web: The Enthralling Structure of the Universe.** The cosmic web is a captivating large-scale

structure of the universe, resembling a spider web of galactic filaments and clusters that separate vast cosmic voids. This complex structure is generated by the gravitational collapse of dark matter, and it is studied using advanced simulations and galaxy surveys. As mere bugs caught in its mesmerizing beauty, we contemplate the universe's interconnectedness and our place within it.

16. **T**he **Speed of Light: The Universe's Limit.** The speed of light is constant in a vacuum, and it is one of the fundamental constants of the universe. This speed is approximately 299,792,458 meters per second and is considered the ultimate speed limit in the universe. According to Einstein's theory of special relativity, nothing can travel faster than light. This means that even if we were to travel at the speed of light, we would never be able to reach the end of the universe. The vastness of the universe is not only due to its size but also the speed of light limit.

17. **T**he **Light Circles the Earth Eight Times in One Second.** The light can travel around the Earth's equator eight times in just one second, covering approximately 24,901 miles (40,075 kilometers). This hard grasping speed is a key factor in many areas of physics, including our comprehension of the cosmos, the behavior of subatomic particles, and the development of communication and navigation technologies.

18. **T**he **Universe is Silent: A Fact That Speaks Volumes.** The universe is remarkably silent due to the lack of a medium for sound waves to travel through space. Despite this, the universe is not entirely devoid of activity, and now scientific

instruments can extract signals, including light, radio waves, and other forms of electromagnetic radiation, and make them audible for the first time.

19. **Simulating the Universe's "Sound": Scientists Translated Data into Audio.** In 2018, a team of scientists used data from the Hubble Space Telescope to capture the first-ever "sound" from the universe. This historic achievement was made possible by converting the data collected from a distant galaxy cluster into a sound frequency that could be heard by humans. Even though sound cannot technically travel through the vacuum of space, this groundbreaking research has opened a new way for scientists to study and explore the universe, all without breaking the laws of physics.

20. **Milky Way's Planetary Abundance: More Planets Than Stars.** The vast expanse of our Milky Way galaxy is home to an incomprehensible number of stars and planets, each contributing to the vast tapestry of cosmic wonder. Recent studies and research have revealed a startling fact about the stellar census of our galaxy: there are more planets than stars. Although the task of accurately counting the number of stars present in the vast expanse of the Milky Way galaxy is daunting, scientists have estimated that the number of stars ranges from 100 billion to an even more mind-boggling 400 billion stars.

21. **Beyond the Moon: Exploring Earth's Other Companions in Space.** The moon is not the only companion of Earth as there are two other celestial bodies orbiting near our planet that are sometimes referred to as moons, although they

are not entirely deserving of the title. One of these is 3753 Cruithne, an asteroid that was discovered in 1986 which orbits the sun. Due to its orbit aligning with Earth's, it appears as though Cruithne is following our planet, and its orbit looks bean-shaped when viewed from Earth. The other body is asteroid 2002 AA29, which also orbits the sun once a year but follows a peculiar horseshoe-shaped path that brings it within 3.7 million miles (5.9 million kilometers) of Earth every 95 years. Because of its proximity to Earth, scientists have suggested that samples should be collected from AA29 and brought back to our planet.

22. **The Multitude of Planetary Systems in the Milky Way: Over 3,200 other "Solar Systems".** Our solar system is just one of many specific planetary systems - a star with planets orbiting around it. Although our planetary system is the only one officially called a "solar system," the reality is that there are numerous other "Solar systems", In fact, astronomers have discovered more than 3,200 stars other than planets orbiting them in their own galaxies.

"

Two possibilities exist either we are alone in the universe, or we are not. Both are equally terrifying.

– Arthur C. Clark

SECTION 2
The Sun: From its Magnetic Field to Its Explosive Activity

23. **The Life Cycle of the Sun: From Main Sequence to Red Giant and the Fate of Our Solar System.** The Sun, a yellow dwarf star that formed from a cloud of gas and dust, is currently in the stable main sequence stage of its life cycle, fusing hydrogen into helium in its core to provide energy and heat to the planets in our Solar System. This process has been ongoing for the past 4.5 billion years and will continue for another 5.5 billion years. However, in about 6.4 billion years, the Sun will have exhausted its hydrogen fuel, leading it to expand and cool, becoming a red giant. This transformation will potentially engulf Mercury, Venus, and Earth. Additionally, the Sun's brightness will increase by 10% in one billion years, causing a moist greenhouse effect on Earth that may threaten life as we know it. Therefore, the Sun's life cycle from main sequence to red giant is crucial for understanding how it sustains life on Earth while highlighting the potential threats to its future.

24. **The Solar Cycle: The Sun's Magnetic Field Affects Our Space Environment.** The solar cycle is a natural cycle that the Sun's magnetic field goes through approximately every 11 years. The Sun's magnetic field generates powerful magnetic fields that cause activity on the Sun's surface, including sunspots. As the magnetic fields change, the amount of activity on the Sun's surface changes too. Scientists track the solar cycle by counting sunspots, with solar minimum being the beginning of the cycle and solar maximum being the peak. Solar activity can affect Earth, causing aurora and impacting radio communications. Scientists work to predict the strength

and duration of solar cycles to help protect satellites and astronauts.

25. **T**he Sun's Address in the Milky Way: A Star on the Edge. The Sun considered the center of our Solar System, is located on the outer edge of the Orion Arm of the Milky Way galaxy. It is approximately 8 kiloparsecs (26,000 light-years) away from the galactic center. The Orion Arm, a region of active star birth, is marked by the presence of massive neighboring stars that could substantially affect the development of planetary systems. Understanding the sun's place in the Milky Way is critical to comprehend our position in the cosmos and the operations of our galaxy. It may be sensible to write down the address in case one day you get lost in space. Though the prospect of becoming lost in space is no laughing matter, at least you'll possess a remarkable postmark to show off.

26. **T**he Powerful Sun: One Million "Earths" Could Fit Inside the Sun. The Sun, the center of our solar system, contains over 99.86% of the total mass in our solar system, making it overwhelmingly dominant. The Sun's diameter is an impressive 864,938.5 miles (1.4 million kilometers), making it approximately 109 times larger than Earth's. The Sun's immense size and gravity have made it the dominant force in the Solar System, determining the orbits and behavior of each of the planets around it and other celestial bodies. It's hard to fathom the vastness of the Sun but to put it into perspective, the Earth could theoretically fit inside the Sun at a rate of 1.3 million times. However, due to the Earth's spherical shape, only around 960,000 Earths could fit within the sun if they were to be nestled.

27. **S**olar Eclipses: Where the Moon Takes Center Stage. A solar eclipse occurs when the moon passes between the sun and the Earth, casting a shadow on the Earth. Solar eclipses can occur only during a new moon and last several minutes. Solar eclipses are relatively rare and can only be observed from a small area on the Earth.

28. **T**he Surprising Science Behind the Color of the Sun. The color of the sun is not as simple as it seems, and it depends on many factors such as the wavelength of the light emitted, the intensity of the light, environmental factors, and the limitations of our eyes and brains. The sun emits energy at all wavelengths from radio to gamma rays, but it emits most of its energy around 500 nm, which is close to blue-green light. The sun's surface temperature, around 5,800K, determines the maximum radiation frequency, which is around the blue-green part of the spectrum. However, our perception of the sun's color is influenced by the atmosphere, which scatters shorter wavelength blue light more efficiently than longer wavelength red light, and the attenuation of all visible light wavelengths passing through our atmosphere. When viewed from space, the sun appears white because it emits strongly in all visible colors, and our eyes perceive a white color due to the integration of these signals into our brains.

29. **T**he Immensity of Our Solar System: 36 billion Times Larger Than Earth. The solar system is much more extensive than just the eight planets orbiting the Sun. It includes the Kuiper Belt, which is a ring of sparsely populated icy bodies beyond Neptune's orbit. Most of these objects are smaller than Pluto. Beyond the Kuiper Belt is the Oort Cloud, which is

a vast spherical shell surrounding our solar system. This shell is made up of icy debris, some of which is mountain-sized. While it has never been directly observed, mathematical models and observations of comets suggest that it exists. To provide some perspective, the distance between Earth and the Sun is about 93 million miles (149 million kilometers), or roughly eight light minutes. The solar system is approximately 36 billion times larger than Earth. The Oort Cloud is located between 5,000 and 100,000 astronomical units from the Sun, where one astronomical unit equals about 93 million miles (149 million kilometers), or the distance from Earth to the Sun. It marks the boundary of the Sun's gravitational influence, where objects can orbit and return to the Sun. In comparison, the heliosphere, which is formed by the solar wind, extends to the termination shock, which is located about 80-100 astronomical units away from the Sun, or roughly 7.5 to 9.3 billion miles (12 to 15 billion kilometers).

30. **The Heart of the Sun: The Core's Nuclear Fusion.** Deep within the Sun lies its powerful core, where most of the nuclear fusion takes place, causing hydrogen atoms to combine and form helium. This incredible process occurs at a temperature of 15 million degrees Celsius, and it powers stars, including the sun, and without it, life on Earth would be impossible. This process creates a vast amount of energy by combining two light atomic nuclei to form a heavier one. The sun's photosphere receives this energy, which is generated by nuclear fusion reactions occurring in its core. Scientists aim to replicate this process in fusion reactors on Earth to meet growing energy needs with clean and efficient energy.

31. **S**olar Flares and Storms: The Explosive Activity of The Sun's Surface. Solar Flares and Storms: Explosions on the Sun's Surface. Solar flares and storms are intense explosive events that occur on the Sun's surface, caused by the release of magnetic energy. These eruptions can cause disturbances in Earth's magnetic field, leading to auroras, disruption of radio communication, and even damage to satellites and power grids. Solar flares occur when magnetic fields on the Sun's surface become twisted and suddenly release their energy in the form of radiation and particles. They are classified based on their X-ray energy output and can range from small C-class flares to the most massive X-class flares. Solar storms, on the other hand, are streams of charged particles that can also disrupt satellite and communication systems on Earth. These events are driven by the Sun's magnetic field, which can become tangled and twisted, releasing huge amounts of energy when it is untangled. While solar flares and storms can pose a threat to Earth's technological systems, they also provide opportunities for studying the Sun's magnetic activity and its effect on space weather.

32. **S**unspots: The Fascinating Magnetic Phenomenon on the Sun's Surface. Sunspots are dark and planet-sized regions of strong magnetic fields on the sun's surface, which can give rise to eruptive phenomena like solar flares and coronal mass ejections. These regions are cooler than their surroundings, hence appearing darker. The level of solar activity during the 11-year solar cycle driven by the sun's magnetic field is indicated by the frequency and intensity of sunspots visible on the surface. Sunspots are a source of fascination for solar observers since they provide a glimpse into the sun's intricate magnetic interior. According to the National Weather Service, the umbra, the central dark region, is about 6,300 degrees Fahrenheit (3,500 degrees Celsius), while the surrounding photosphere is about 10,000 Fahrenheit (5,500 degrees Celsius).

33. **The Sun's Halo: The Formation of the Solar Crown.** Solar halos are observable circular or semicircular arcs of light around the Sun, caused by the refraction and reflection of sunlight by ice crystals in the atmosphere. These halos can be observed in many parts of the world, especially in the polar and cold regions, and are best seen at dawn or dusk. While they can be breathtakingly beautiful, it should be noted that staring at them for too long can cause eye damage.

34. **Exploding with Power: Coronal Mass Ejections from the Sun's Fiery Corona.** Coronal mass ejections, a violent solar phenomenon, occur when the sun's magnetic activity produces huge clouds of plasma and magnetic fields that are expelled from the corona. These explosions can cause geomagnetic storms that create stunning auroras and affect satellites and power grids. To study these events, scientists use instruments like the LASCO on the SOHO satellite. Interestingly, the sun's corona can reach temperatures that are 500 times hotter than the photosphere, ranging from 1.8 million degrees to 3.6 million degrees Fahrenheit (1 to 2 million degrees Celsius).

35. **Only Two Billionth of the Sun's Energy Reaches Earth.** Despite being relatively close to the sun, only a tiny fraction of the sun's energy reaches the Earth. This is because the sun emits a wide range of wavelengths, super Jupiter's atmosphere, and magnetic field block or deflect most of them.

Can you believe it? We only get a measly two billionth of the sun's energy! And yet, we astoundingly still manage to power our homes, cars, and phones with this infinitesimal amount. I mean, talk about efficiency! It's like using a single grape to make a whole exquisite bottle of wine.

36. **Twister on the Sun: Solar Tornadoes.** Solar tornadoes are among the most impressive events in the solar system. These tornadoes are caused by magnetic structures that have a spiral shape and arise from the surface of the Sun. The magnetic structures are anchored at both ends to the Sun's surface, which causes them to rotate in a swirling manner similar to a tornado. Some of the largest structures ever seen, the width of these structures can reach 100,000 miles (160,934 kilometers). Plasma discharges move with the spiral-shaped magnetic field as they take place inside the structure. Solar tornadoes are beautiful and important solar system phenomena that are created because of this intriguing process.

37. **Solar Filaments: The Longest Structures in the Solar System**. Solar filaments, also known as prominences, are undeniably one of the most intriguing and magnificent phenomena present in the solar system. These elongated structures, stretching over hundreds of thousands of miles (or kilometers), are crafted from plasma, which is an intensely hot gas comprising charged particles. What is truly remarkable about these filaments is the fact that they are anchored in place by the Sun's magnetic field, which gives them their elongated shape. Perhaps the most fascinating aspect of solar filaments is their exceptional stability, as they can persist for weeks, and sometimes even months, before finally erupting or fading away. When a filament does erupt, it unleashes an enormous

quantity of matter and energy into space, causing a coronal mass ejection (CME). These CMEs can trigger disturbances in the Earth's magnetic field and have the potential to inflict damage on electrical and communication systems. Despite the potential hazards, the study of solar filaments is critical for comprehending the behavior of the Sun and forecasting space weather events. These captivating structures function as a striking reminder of the incredible power and intricacy of our nearest star, showcasing its majestic beauty, which is unparalleled in the cosmos.

38. **The Force of the Sun's Gravity Holding the Solar System Together.** The Sun's gravitational force is responsible for holding the Solar System stable by extending its impact to all planets and celestial bodies within its reach and preventing them from careening off into the endless void of space. Even though the Sun's gravitational pull becomes less powerful as we go farther away from it, it is still strong enough to keep the entire Solar System in a steady orbit around the Sun.

39. **The Sun's Closest Companion: Proxima Centauri.** Proxima Centauri, lying only 4.24 light-years away, is the Sun's closest companion star and a member of the Alpha Centauri triple star system. This remarkable discovery not only unveils the proximity of binary star systems like the Sun and Proxima Centauri but also emphasizes their prevalence in the vast universe.

40. **The Sun's Secret Emissary: The Rapid Solar Wind.** The solar wind, a high-speed stream of charged particles consisting mainly of electrons and protons, continuously flows out from

the Sun's outer atmosphere, the corona, and expands into space at breakneck speeds ranging from 671,080.56 miles per hour to over 1,790,861.12 miles per hour (300 km/s to over 800 km/s). The solar wind carries the Sun's magnetic field with it and can significantly impact the environment of celestial bodies it encounters. The solar wind is responsible for auroras on Earth and disruptions in communication and navigation systems. This powerful stream of particles can also influence the formation and evolution of comets, asteroids, and other small bodies in the Solar System, shaping our cosmic neighborhood in extraordinary ways.

41.   **A**stonishing Speed: The Velocity of Our Solar System. The solar system is hurtling through the vast expanse of the galaxy at an astonishing speed of 514,000 miles per hour (828,000 kilometers per hour) as a result of the Sun's orbit around the center of the Milky Way galaxy.

42.   **T**he Heliosphere: The Bubble of the Solar System. The Heliosphere is a vast region of space surrounding our Solar System, extending to several hundred astronomical units (AU) from the Sun. It is the boundary between the solar wind and the interstellar medium. The Heliosphere is shaped like a bubble; it is thought to be created by the solar wind, which pushes back against the interstellar medium. The Voyager 1 and Voyager 2 spacecraft have reached the Heliosphere's edge, known as the Heliopause, and have provided scientists with valuable data on this crucial boundary. The Heliosphere protects us from harmful cosmic rays.

"

*Once you can accept
the universe as matter
expanding into nothing that
is something, wearing stripes
with plaid comes easy.*

– Albert Einstein

SECTION 3
The Fascinating World of Planetary Formation

43. **T**he **Formation of Planets: From Dust to Planets.** Planets form from the dense disk of gas and dust surrounding young stars, which contain elements like carbon and iron that contribute to the formation of planetary systems. During the T Tauri phase, the star is emitting hot winds dominated by protons and helium atoms, while dust particles collide and clump together, forming pebbles and rocks. Gas helps these particles stick together and become planetesimals, the building blocks of planets. In colder regions, ice fragments combine with dust to form icy giants like Jupiter and Saturn, while warmer regions closer to the star form rocky planets like Earth. Some break apart, but others continue to grow and eventually form planets.

44. **T**he **Ice Giants of the Solar System: Uranus and Neptune.** The celestial twins of the outer solar system share a striking resemblance in their composition. Predominantly composed of ice such as water, ammonia, and methane, these two planets have earned the classification of "ice giants" by scientists. Their similarity in composition and structure offers valuable insights into the formation and evolution of our solar system.

45. **T**he **Formation and Age of Terrestrial Planets: Accretion and Radioactive Elements.** The process of creating terrestrial planets is called accretion, which involves the collision and fusion of small particles that gradually form larger and larger bodies, eventually resulting in the formation of a complete planet. Scientists can estimate the age of a planet by measuring the quantity of radioactive elements present in rocks on its surface, providing important insights into the history and evolution of our solar system.

46. **The Solar System's Inner Circle: The Terrestrial Planets.** The solar system's inner planets, including Mercury, Venus, Earth, and Mars, are known as terrestrial planets. These planets are composed mainly of rock and metal and have solid surfaces. They are also relatively close to the sun and much smaller than the outer gas giants. The terrestrial planets have various geological features, and Earth is the only one known to support life.

47. **Gas Giants: Rapid Formation and Habitable Planet Implications.** The process of gas giant planet formation is still a subject of scientific curiosity and research. These planets are composed of a large solid core surrounded by a significant amount of hydrogen and helium gas. In about 200,000 years, the most common gas giants in the universe, such as Jupiter and Saturn, can form quite rapidly by incorporating nearby icy bodies composed of drifting pebbles from the outer parts of young planetary systems. Understanding how gas giants form is important not only for our own solar system but also for understanding the creation of habitable planets in other solar systems.

48. **The Solar System's Outer Giants: The Gas Planets.** The solar system's outer planets, including Jupiter, Saturn, Uranus, and Neptune, are known as gas giants. These planets are much larger than the inner terrestrial planets and are composed primarily of gas and ice. They also have much longer orbits and

are much farther from the sun. The gas giants have fascinating features such as Jupiter's Great Red Spot, Saturn's rings, and Uranus' unique tilt. These planets have multiple moons, some with interesting geological features, such as active volcanoes on Jupiter's moon Io and subsurface oceans on Saturn's moon Enceladus.

"

Look up at the stars and not down at your feet. Try to make sense of what you see and wonder about what makes the universe exist. Be curious.

– Stephen Hawking

SECTION 4
The Incredible Shrinking Planet: Mercury

49.   **The Incredible Shrinking Planet: Mercury.** The smallest planet in the solar system (excluding the dwarf planet Pluto) is also the second densest after Earth. But it's not just small; it is also getting smaller and thicker over time. This shrinkage is evident in the planet's landforms, shaped by billions of years of contraction. In the 19th century, geologists already hypothesized that some of Earth's most dramatic landforms emerged as the planet shrank.

50.   **Fastest Planet in the Solar System: Mercury**. Mercury, the swiftest planet in our solar system, is named after the Roman god of messengers, and for good reason. With an impressive orbital velocity of about 107,000 miles per hour (172,200 kilometers per hour), it completes a full orbit around the sun in just 88 Earth days. The planet, known to ancient people thousands of years ago, has no moons and is also the smallest planet in the solar system. As the closest planet to the sun, it experiences scorching temperatures, reaching up to 800 degrees Fahrenheit (430 degrees Celsius) during the day and dropping to a negative 290 degrees Fahrenheit (-180 degrees Celsius) at night. Despite its small size, Mercury's rapid orbit and proximity to the sun make it a fascinating and vital subject of study for scientists seeking to understand our solar system.

51.   **Mercury's Tail: A Spectacular Solar Wind Phenomenon.** Mercury's tail is a magnificent phenomenon that occurs when the solar wind pushes sodium atoms from the planet's surface, creating a stunning tail akin to that of a comet. This incredible

sight can be viewed using a telescope and a special filter that captures a long-exposure photo, leaving onlookers in absolute amazement. The tail is brightest within 16 days of the planet's perihelion and interested observers can use the Sky Tonight app to find a good day to view it.

52. **M**ercury: The Atmosphere-Challenged Planet. Mercury, the closest planet to the sun, has a very thin atmosphere known as an exosphere. This atmosphere is composed mostly of atoms weathered by the solar wind and meteoroids. However, it is not capable of supporting life and has extremely low surface pressure. The elements that make up this atmosphere are mainly helium and sodium. Additionally, due to the planet's small size and lack of a protective magnetic field, its atmosphere is constantly lost to space, making it a challenging environment for spacecraft attempting to land on its surface.

53. **M**ercury's Time Warp: Where a Day is Longer than a Year. Mercury experiences a unique time dilation due to its close proximity to the sun and spin-orbit resonance. The planet completes one rotation on its axis, or rotational period, in about 59 Earth days. It takes 88 Earth days for the planet to complete one orbit around the sun. However, the length of a day on Mercury, defined as the time between two successive sunrises, is 176 Earth days. This peculiar time warp is caused by the planet's 3:2 spin-orbit resonance, meaning it completes three rotations on its axis for every two orbits around the sun. As a result, a year on the planet is shorter than a day on the planet. Additionally, Mercury's proximity to the sun also causes time dilation, a consequence of Einstein's theory of general relativity. This states that time passes slower in stronger gravitational fields. Therefore, a clock on Mercury would run slower when viewed from Earth. However, this effect is not perceivable for humans or any macroscopic objects.

54. **M**ercury's **Extreme Temperatures**. Mercury is the closest planet to the sun, so it experiences the most extreme temperature fluctuations in the solar system. Temperatures can reach as high as 800 degrees Fahrenheit (427 degrees Celsius) during the day and drop as low as -290 degrees Fahrenheit (-179 degrees Celsius) at night. This makes Mercury's surface hot enough to melt lead and cold enough to freeze oxygen.

55. **M**ercury's **Surface: Unique Landforms and Volcanic Activity.** Mercury's surface is characterized by unique landforms and volcanic activity, as well as a high density due to its iron-sulfur core. Although it appears similar to the Moon's surface, Mercury has distinct differences, such as low iron and high sulfur surface compositions, lobate scarps indicating crustal contraction, and features like the 1,300 km diameter Caloris Basin, which was flooded by lava after impact, and the "wheel and spoke" patterned Rembrandt Basin. Impact craters on Mercury, which are similar to those on the Moon, are deeper and more complex due to the planet's high surface gravity, causing ejected material to travel shorter distances and resulting in secondary craters. Multiring basins and smaller craters with central peaks and flattened floors are also common on Mercury, possibly due to a lower content of volatile materials or higher impact velocities.

56. **M**ercury's Caloris Crater: A Young and Unique Feature with Intricate Patterns. The surface of Mercury features a remarkable geological feature called the Caloris impact basin, measuring 1,550 km in diameter and surrounded by distinct terrains. The interior of the basin comprises smooth plains with intricate radial and concentric patterns, while the basin rim consists of irregular mountain blocks towering up to 3 km in height. The outer region of the basin contains linear, radial ridges, and valleys filled with plains and volcanic vents. Although scientists still debate the formation of the plains in the annulus region, evidence suggests that it may be a result of basin ejecta and volcanism. The Caloris Basin is considered to be one of the youngest basins on Mercury and may have formed approximately 3.9 billion years ago, around the same time as the last giant basins on the Moon. Another notable feature is the Raditladi basin, which is smaller in size but boasts a prominent interior ring and may have formed more recently.

"

When we gaze at the stars, we witness the whispers of eternity. In their twinkling light, we find solace, for they remind us that our fleeting lives are but a part of a vast cosmic dance.

— Martin Heidegger

SECTION 5
Venus: The Scorching Jewel of Our Solar System

57. **The Hottest Planet in the Solar System: The Scorching and Toxic Venusian Environment.** The atmosphere of Venus is a toxic and scorching environment unlike any other in our Solar System. It is composed primarily of carbon dioxide, which makes up 96.5% of the atmosphere and is thick enough to create a crushing pressure at the surface over 90 times greater than Earth's. The thick atmosphere also creates a greenhouse effect, raising temperatures on the surface to a scorching 863.6 degrees Fahrenheit (462 degrees Celsius). The clouds in the Venusian atmosphere are made of sulfuric acid, giving the planet its distinctive yellowish hue. The Venusian atmosphere is an actual inferno, a hostile environment unsuitable for any known form of life. It would be incumbent for any aspiring interplanetary traveler to consider the hazards of a Venusian excursion with the utmost care.

58. **The Radiant Venus: The Second Brightest Object in the Night Sky.** Venus is the second brightest natural object in the night sky after the Moon; its atmosphere is covered in sulfuric acid clouds that make it reflective and shiny, obscuring the planet's surface. With temperatures reaching 9,000 degrees Fahrenheit near the core (4,982 degrees Celsius), Venus is a scorching and radiant planet in the night sky. Venus earns the title of the 'Mirage Planet.'

59. **Venus: The Most Volcanic Planet in The Solar System.** Venus is home to the most volcanoes in the Solar System, with over 1,600 major volcanoes or volcanic features. There may be even more than 100,000 or more than 1 million smaller volcanoes. Although recent data suggest that the planet still has active volcanoes, the evidence must be more conclusive. Some recent findings indicate that a volcanic peak more than a mile high on Venus may still be active, potentially shedding light on hotly debated findings suggesting that life exists on the hellish planet. If Venus were a human personification, one might describe it as gothic, with its ominous, dangerous atmosphere and enigmatic volcanic activity.

60. **The Slow and Disorienting Rotation of Venus: A Backward Year and a Day Longer Than Its Year.** Venus has a rotation period of about 243 Earth days, which is equivalent to almost eight Earth months for a Venusian "day." Venus rotates on its axis in the opposite direction to its orbit around the sun, resulting in the sun appearing to rise in the west and set in the east. Due to this slow rotation, one Venusian "year" is slightly shorter than one Venusian "day," and the duration of a sunrise-to-sunset on Venus is about 117 Earth days. This experience can be quite disorienting for an observer on the planet's surface.

"

*I don't pretend to understand
the universe—it's much bigger
than I am.*

– Albert Einstein

SECTION 6 Earth:
A Small, Blue Marble in the Vast Universe

61. **E**arth's Mass Compared to the Moon: A Stunning Difference of 81 Times. The Earth, our home planet, has a much greater mass than its natural satellite, the Moon. The Earth has a mass of approximately 5.9 x 10^24 kg, while the Moon has a mass of roughly 7.3 x 10^22 kg. This difference in mass is likely due to the moon's smaller size and lighter composition. This heavyweight of the Earth causes the Moon to orbit around it, which is also why we have a stronger gravitational pull on Earth than on the Moon.

62. **T**he Round Truth: Earth's Spherical Shape. The Earth is not flat, as some ancient cultures believed, but is an oblate spheroid. It is slightly flattened at the poles and bulging at the equator due to the Earth's rotation and the centrifugal force it creates. Pythagoras and Galileo Galilei confirmed this shape in their experiments, and today it is proven by satellite imagery and measurements of the Earth's gravity field. The curved horizon is further proof of the Earth's round shape.

63. **T**he Inner Earth's Core: A Hot Heart. The core of the Earth is its deepest layer and is divided into two parts: the liquid outer core and the solid inner core. Iron and nickel make up most of this extremely hot interior, which reaches temperatures of about 9,932 degrees Fahrenheit (5,500 degrees Celsius), while iron and nickel also make up the outer core, which reaches temperatures of about 7,232 degrees Fahrenheit. (4,000 degrees Celsius). The magnetic field generated by the Earth's

core is a powerful force that propels plate tectonics and shapes our planet's geological surface.

64. **E**arth's Unique Qualities: Internal Heat, Liquid Water. The Earth has two unique features that differentiate it from other terrestrial planets - an abundance of internal heat and liquid water. The Earth's size and mass have allowed it to hold onto its internal heat, which keeps the surface deformable and prevents the lithosphere from becoming too rigid. In comparison, Mars is smaller and has cooled much faster, causing its lithosphere to become too rigid to break into plates. While Venus is similar in size to Earth, it lacks moving plates because it does not have interior water to lubricate them. Although tectonic deformation occurs in the outer layers of Venus and Mars, they are considered one-plate planets because their outer layers are not broken up into plates. However, Jupiter's moon Europa may experience plate tectonics due to its shell of cold, brittle ice floating atop a warmer, fluid ice layer, with evidence of water upwelling to the surface.

65. **G**ravity's Surprise: Earth's Gravity Isn't Uniform. If the Earth were a flawless sphere, the gravitational force would be uniform everywhere. However, due to the planet's uneven surface, the influence of gravity changes because of factors like the movement of the tectonic plates beneath the Earth's crust, ice displacement, and water flow. These fluctuations are referred to as gravity anomalies. When there is a mountain range like the Himalayas, it creates a positive gravity anomaly where gravity is more forceful than on an unblemished, level planet. Conversely, ocean trenches or land depressions caused by ancient glaciers create negative gravity anomalies. NASA's GRACE (Gravity Recovery and Climate Experiment) mission is

revolutionizing the mapping of the Earth's gravitational field by observing it from space.

66. **Shielding the Planet: The Magnetic Field of Earth and its Fascinating Properties.** The magnetic field of the Earth, also known as the geomagnetic field or magnetosphere, protects our planet from the solar wind's charged particles. It is generated by electric currents from various parts of the Earth and has two poles; the magnetic north pole, which is close to the geographic north pole, and the magnetic south pole, which is close to the geographic south pole. This arrangement makes the compass useful for navigation. The field is infinite but weakens with distance from its source. The Earth's magnetic field extends tens of thousands of kilometers into space, creating the magnetosphere. The Earth's magnetic poles are not stationary and wander over time, and the magnetic field also flips irregularly, reversing the north and south magnetic poles. The magnetic field is at least 3.5 billion years old, according to a paleomagnetic study of Australian red dacite and pillow basalt.

67. **The Earth's Atmosphere: A Heterogeneous Mixture of Gases and Conditions Supporting Life.** The Earth's atmosphere is an intricate and diverse blend of nitrogen (78%), oxygen (21%), and trace gases like carbon dioxide and argon. This heterogeneous mixture creates an environment that sustains life, shielding us from harmful solar radiation and maintaining temperatures that allow for the existence of liquid water. The atmosphere also plays a crucial role in regulating the planet's climate through mechanisms such as the greenhouse effect and precipitation formation and participates in the water cycle by facilitating the evaporation, ascent, and descent of water vapor.

68. **T**he 6,200 Miles of Earth's Atmosphere in Five Distinct Layers. The Earth's atmosphere is composed of five distinct layers, each with different thermal characteristics, chemical composition, movement, and density. The outermost layer, the exosphere, extends from about 375 miles to 6,200 miles (603.5 kilometers to 9,977 kilometers approx.) above the Earth, where satellites orbit, and molecules escape into space. Below that is the thermosphere, which absorbs high-energy radiation from the sun and becomes denser as one descends, with temperatures ranging from (-184 degrees Fahrenheit to 3,600 degrees Fahrenheit (-120 degrees Celsius to 1,982 degrees Celsius). The mesosphere, located between 31 miles to 53 miles above the Earth's surface, is where meteors burn up upon entry into the atmosphere. The stratosphere holds 19% of the atmosphere's gases but very little water vapor, and the troposphere, which is the layer closest to the Earth's surface, is where almost all weather occurs. The height and temperature of each layer vary depending on its location in the atmosphere.

69. **T**he Vital Role of the Ozone Layer in Protecting Life on Earth. The ozone layer is a vital component of the Earth's atmosphere, located within the second layer called the stratosphere. The stratosphere is a protective mass of gases surrounding our planet, and it gets its name from its stratified or layered structure. As altitude increases, the stratosphere becomes warmer due to the absorption of ultraviolet radiation by ozone gases in the upper layers. Despite being a trace gas in the atmosphere with only about three molecules for every 10 million molecules of air, the ozone layer plays a crucial role in shielding life on Earth. It acts like a sponge, absorbing harmful radiation from the sun that could otherwise damage living organisms. Although some of the sun's radiation is necessary for life, too much of it can be detrimental, and the ozone layer acts as a protective barrier against such harmful effects.

70. **R**acing **through Space: The Mind-Boggling Speeds of Earth's Journey Through the Universe.** The Earth moves at an average speed of 67 miles per hour (30 km/s) as it orbits around the Sun, and it rotates around its axis at a speed of 1.19 miles per hour (0.5 km/s) at the equator. The solar system travels at 492 miles per hour (220 km/s) as it orbits around the center of the galaxy, while neighboring galaxies move towards the Great Attractor at a speed of almost 2,237 miles per hour (1,000 km/s). The cosmic microwave background radiation (CBR) provides a universal frame of reference, and the Earth moves at a speed of 872,405 miles per hour (390 km/s) in the direction of the Leo constellation relative to the CBR.

71. **W**ithout **the Atmosphere, Earth would be a Frozen Wasteland.** Without the atmosphere, Earth would be a frozen wasteland. The atmosphere is crucial for the survival of life on Earth. Without it, the surface would be a frozen wasteland with extreme temperatures ranging from -374 degrees Fahrenheit (-190 degrees Celsius) to 230 degrees Fahrenheit (110 degrees Celsius), making it uninhabitable for all types of life as we know it.

72. **T**ime **Warp: The Slowing of Earth's Rotation.** The Earth's rotation is slowing down due to friction effects caused by the Moon's tides. Every century, this process adds about 2.3 milliseconds to the length of each day. When Earth formed 4.6 billion years ago, a day lasted only six hours. Time is relative, and even the Earth takes its time.

"

*The cosmos is a riddle,
an enigma that invites
us to contemplate the
immeasurable depths of its
mysteries. It is in our quest
to unravel its secrets that we
uncover the hidden truths of
our own souls.*

– Simone de Beauvoir

SECTION 7
Mars: Our Potential Alternative Home

73. **W**ater on Mars: The Key to Unlocking the Mystery of Life Beyond Earth. Water on Mars is a fascinating topic for scientists, as it holds the potential for discovering life beyond our planet. Evidence has been found for water in both its frozen and liquid forms. Mars has an abundance of water ice, with enough to fill Lake Superior, and more than 5 million km3 of ice have been detected at or near the surface of the planet. However, the atmospheric pressure is too low for large standing bodies of liquid water to exist today. Instead, liquid water is thought to be located in subsurface reservoirs accessible through advanced drilling techniques. Despite this, there is evidence of past water on Mars, including enormous outflow channels, ancient river valley networks, and the remains of an ancient freshwater lake, such as those found in Gale Crater, which may have once been a hospitable environment for microbial life. The discovery of water on Mars not only reveals the planet's potential habitability but also opens up exciting new possibilities for exploring the possibility of life beyond Earth.

74. **M**ars: Once an Earth-Like Planet with Potential for Life. Mars may have once been similar to Earth billions of years ago, with a dense atmosphere, magnetic field, and possibly even liquid water on its surface. This created an environment that could have supported life. NASA has collected evidence supporting these theories, including the presence of water on Mars in the past. Recent studies suggest that Mars may have had enough water to form a global ocean up to 300 meters deep, making it the first planet in the Solar System with the potential to sustain life.

75. **M**ars' **Thin Atmosphere: 95% Carbon Dioxide and Less than 1% Oxygen.** The atmosphere is composed of 95.32% carbon dioxide, 2.7% nitrogen, 1.6% argon, and 0.13% oxygen. The atmospheric pressure at the surface is 6.35 mbar, which is over 100 times less than Earth's. Water exists on Mars, but due to the thin atmosphere, it cannot last long on the surface in a liquid state. However, water ice exists below the surface of the polar regions, and seasonal briny water flows down hillsides and crater walls. Despite the thin atmosphere, Mars experiences dynamic weather events like dust storms and snow. In the past, Mars had a thick atmosphere that could have supported liquid water on the surface. For human exploration efforts, generating oxygen from the Martian atmosphere is crucial, as humans cannot breathe the current composition.

76. **M**ars: **Fluctuating Temperatures and Thin Atmosphere.** Mars has colder weather than Earth due to its thin atmosphere and distance from the sun. The average temperature on Mars is around -80 degrees Fahrenheit (-60 degrees Celsius) with daily temperature fluctuations ranging from -195 degrees Fahrenheit (-125 degrees Celsius), at the poles during winter to a pleasant 70 degrees Fahrenheit (20 degrees Celsius) at the equator during midday.

77. **M**ars' **Equatorial Paradox: From Heat to Freeze in a Day.** Mars is characterized by its extreme temperature fluctuations, with daily temperature differences reaching 100 degrees Fahrenheit (38 degrees Celsius). During the day, the Martian equator can reach a warm 68 degrees Fahrenheit yet plummet to an icy -99° degrees Fahrenheit (20 degrees Celsius to -73 degrees Celsius) at night. This creates a temperature paradox, spring at the feet and winter at the head in just one day. Mars is the perfect destination for those who can't decide between a tropical and a winter holiday.

78. **The Martian Wind: A Red Planet's Obstacle.** Mars poses a challenge for spacecraft landings due to its strong winds that can reach up to 60 mph (96.56 kph) and vary in color from orange to red. These winds can interfere with the landing system's proper functioning, creating difficulties for scientists trying to successfully land a spacecraft on Mars. However, the wind has a beneficial effect as well; it can remove dust from the solar panels, providing more power to the mission.

79. **Massive Dust Storms Ravage Mars: Impacting NASA's Explorers and Monitoring Orbiters.** Mars experiences large and long-lasting dust storms, which can cover the entire planet. In January 2022, a dust storm almost twice the size of the United States covered the southern hemisphere of Mars, affecting NASA's explorers on the surface, including the Insight lander and the Ingenuity helicopter. NASA orbiters such as the Mars Reconnaissance Orbiter, MAVEN, and Odyssey monitor and relay data back to Earth from the rovers and landers on the ground. During the dust storm, Odyssey helped InSight by overcoming its technical issues.

80. **Largest Volcano in The Solar System: Olympus Mons.** Olympus Mons, the colossal volcano situated on the red planet, is a true giant among the natural wonders of our

solar system. Standing at a staggering height of 13.05 miles (21 kilometers), it is not only the largest volcano in the solar system but also three times the size of Mount Everest, the tallest peak on Earth. This massive volcano is comparable in size to the state of Arizona, and its slopes are gentle enough to ski on. It is a testament to the raw power of nature and serves as a reminder of the incredible geological forces that shape our planet and its neighboring worlds.

81. **The Largest Canyon in the Solar System: Valles Marineris.** Valles Marineris, also known as the Mariner Valley, is a massive canyon along the Martian equator. It is over 2,485.48 miles (4,000 kilometers) long, up to 4.35 miles (7 kilometers) deep, and as wide as 372.82 miles (600 kilometers). The canyon is thought to have formed from the stretching and cracking of the Martian crust.

82. **The Gale Crater: Craters on the Red Planet.** On the Red Planet, there's a crater so massive that it makes you feel like you're looking at something truly out of this world. Known as the Gale Crater, it stretches out over 154 km in diameter and sits in the southern hemisphere of Mars. It's named after the astronomer Walter Frederick Gale, who gazed upon Mars in the late 1800s. The Gale Crater is a hotbed of scientific fascination because it offers tantalizing clues to the history of water on Mars.

83. **Marsquakes: Seismic Activity on the Red Planet.** A marsquake is similar to an earthquake on Earth, caused by the sudden release of energy within the planet's interior, possibly from plate tectonics or hotspots. Detecting and analyzing

marsquakes can provide information about the interior structure of Mars and its volcanoes. Mars has been seismically active in the past, as evidenced by magnetic striping and Valles Marineris, which may be an ancient Martian strike-slip fault. In 2019, marsquakes were definitively observed, and the first confirmed seismic event from Valles Marineris was detected by InSight in August 2021, a quake with a magnitude of 4.2.

84.   **M**artian Sunsets: The Mesmerizing Blue Sky. As one might expect, the sky on Mars is occasionally red. However, due to Mars's thin atmosphere, which is less than 1% of the density of Earth, the sunset on the red planet is not always red but also blue. The sunset on Mars is blue, unlike on Earth, where it's predominantly blue during the day and orange-red during dawn and dusk. This color difference is due to the Martian atmosphere being dominated by large-sized dust particles that cause 'Mie scattering', which filters out red light and only captures blue light from the sun to reach our eyes, making the Martian sky a mesmerizing shade of blue that would make even the most hardened space adventurer stop and stare.

85.   **T**he Allan Hill Meteorite: A Controversial Martian Rock. The Allan Hill Meteorite, also known as ALH 84001, is a Martian meteorite that was found in Antarctica in 1984 by American meteorite hunters from the ANSMET project. It belongs to a group of meteorites called shergottite-nakhlite-chassignite and is believed to have originated from Mars. In 1996, claims were made that fossilized bacteria were found in the meteorite, leading to widespread media attention and discussion of life on Mars. However, the scientific community ultimately rejected these claims, though the discovery remains significant in the field of astrobiology.

"

We are not just figuratively,
but literally stardust.

– Neil deGrasse Tyson

SECTION 8
Jupiter: The Gigantic World of Gas and Storms

86. **J**upiter: **The Magnificent Giant of our Solar System**. The largest and fifth planet in our solar system has a radius of 43,440.7 miles (69,911 kilometers) and is 11 times wider than Earth. It is so big that all the other planets in the solar system could fit inside it; more than 1,300 Earths could fit inside Jupiter. The planet, known as a gas giant, is primarily formed of hydrogen and helium, with traces of other gases. Jupiter boasts a diameter of roughly 86,881 miles (139,822 kilometers) and a mass that is more than 2.5 times greater than that of all the other planets in the solar system combined. And it is approximately 318 times more massive than Earth specifically.

87. **J**upiter's **Size Makes the Sun Orbit Around It: The Secrets of the Solar System's Barycenter.** The barycenter is the common center of mass between two or more objects, including planets and stars. It is the point at which the mass of an object is balanced. In our solar system, the barycenter between the sun and Earth is very close to the center of the sun because the sun is much more massive than Earth. However, the barycenter between the sun and Jupiter is just outside the sun's surface because Jupiter is much larger than Earth. In this case, the barycenter of both the Sun and Jupiter orbit around their common center mass. Jupiter doesn't orbit around the sun due to its magnitude. However, the barycenter of the entire solar system changes position as the planets move in their orbits, causing the sun to wobble. Astronomers use the wobbling of stars to find planets beyond our solar system.

88. **J**upiter **Belts and Zones: A Unique Characteristic of the Gas Giant.** Jupiter is distinctive for its belts and zones, which are white and reddish bands of clouds that wrap around the planet. NASA's Juno mission has discovered that the prominent bands on Jupiter's surface, which are characterized by powerful east-west winds or jet streams, reach depths of almost 2,000 miles (3,200 kilometers). These jet streams create weather patterns like cyclones and anticyclones on the planet's surface, adding to Jupiter's unique characteristics.

89. **J**upiter's **Magnetic Monster: The Giant Planet's Powerful Field.** Jupiter, the largest planet in our Solar System, boasts a magnetic field that is a force to be reckoned with. It is 20,000 times stronger than Earth's due to its rapid rotation. This powerful field creates stunning auroras on Jupiter's poles, intense radiation belts, and strong winds, leading to the formation of the Great Red Spot, a massive storm that has been raging for at least 350 years.

90. **J**upiter's **Great Red Spot: A Cosmic Storm for Centuries.** Jupiter, the largest planet in our solar system, is known for its colossal magnetic field that is 20,000 times stronger than Earth's, thanks to its rapid rotation. This powerful magnetic field is responsible for creating stunning auroras on Jupiter's poles, intense radiation belts, and the iconic Great Red Spot. With a size larger than that of Earth, the Red Spot is a massive storm system that has been raging for at least 350 years and possibly as long as 350- 700 years. Despite being studied for centuries, scientists are still trying to unravel the mystery of how it can maintain its size and intensity for so long. The Red Spot is classified as a giant anticyclone, much like the polar vortices on Earth, but it has winds that can reach up to approximately 385.25 miles per hour (620 kilometers per hour) and is visible from Earth with a telescope.

91. **A** Shrinking Torment: The Changing Face of Jupiter's Great Red Spot. The Great Red Spot, a massive storm on Jupiter, has been fascinating astronomers for over a hundred years. Despite extensive research, scientists are still puzzled about how it has persisted and remained active for so long. This immense anticyclone has some resemblance to the polar vortices found on Earth, but with winds that can reach up to 385 miles per hour (620 kilometers per hour). The storm is so powerful that it can be observed from Earth through a telescope, making it a remarkable and awe-inspiring natural phenomenon.

92. **J**upiter's Impact on Planetary Systems: Protector or Destroyer? Jupiter is a large planet that can migrate from its original formation location and move closer to its star, which can result in the consumption or flinging out of smaller, rocky planets from the star system due to its strong gravity. However, if Jupiter remains far from its star, it can act as a gatekeeper to the planetary system, keeping the orbits of the inner planets stable, and preventing climate shifts. Jupiter also prevents asteroids in the solar system's asteroid belt from combining into a planet. Moreover, Jupiter can influence the orbits of small bodies that come close to it by sending them on long orbits that take hundreds or thousands of years to return. This is how comets form the Oort cloud due to Jupiter's influence. While Jupiter often protects Earth by deflecting comets and asteroids, it can also send objects on a collision course toward the inner planets, which could be disastrous, as

seen in the impact that caused the extinction of the dinosaurs. Thus, Jupiter's powerful gravitational influence can be both a protector and a destroyer of planetary systems.

93. **T**he Second Planet with More Moons: Jupiter Galilean Satellites. To date, Jupiter has 95 confirmed moons, four of which were first discovered by Galileo in 1610 and are now known as the Galilean satellites: Io, Europa, Ganymede, and Callisto. These four moons are some of the largest objects in the solar system, second only to the Sun and the eight planets. However, this count does not include the numerous small meter-sized moons that are believed to have been ejected from the inner moons, as well as the hundreds of irregular outer moons, which are several kilometers in size and have only been briefly observed with telescopes.

"

Space, the silent canvas of infinite possibilities, beckons us to explore its depths. As we venture into the unknown, we come face to face with the profound questions of our existence, discovering that the cosmos holds both the answers and the unfathomable beauty of the questions themselves.

– Karl Jaspers

SECTION 9
Saturn: The Ringed Wonder of the Solar System

94. **The Lighter Planet in The Solar System: Less Dense Than Water.** Saturn, the second largest planet in our Milky Way galaxy, is known for its low density and gaseous composition. Its density is less than water's, making it float if placed in a liquid of the same density. This unique characteristic sets it apart from the other planets in our solar system, making it an intriguing study object for scientists and astrologers.

95. **Saturn's Hexagonal Storm: A Cosmic Phenomenon.** Saturn's North Pole is home to a strange hexagonal-shaped storm, with each side measuring over 13,800 km long. Scientists are still trying to understand how this hexagonal shape is maintained and what causes it. It may be caused by a jet stream that flows around the pole, but more research is needed to understand this unique feature of Saturn's atmosphere.

96. **The Great White Spot: Saturn's Enormous and Mysterious Storm.** The Great White Spot, also referred to as the Great White Oval, is a periodic storm on Saturn that can be seen from Earth due to its distinct white appearance. These storms can have a width of several thousand kilometers and are named based on their similarity to Jupiter's Great Red Spot. The 2010-11 instance of the storm, also known as the Northern Electrostatic Disturbance, was monitored by the Cassini orbiter, which detected an increase in radio and plasma interference. Data collected by Cassini revealed changes in the storm's clouds, including a reduction of acetylene, a rise in phosphine, and an abnormal temperature decrease in the center of the storm. In 2012, after the visible aspects of the storm had died down, two hotspots merged and emitted a burst of heat and ethylene.

97. **S**aturn Reigns Supreme: The Planet with the Most Moons in Our Solar System. Saturn is known to have a great number of moons in the solar system, with more than 117 confirmed and more under investigation. These moons have diverse characteristics and sizes, with names such as Titan, Rhea, and Enceladus. While Jupiter is renowned for its vast collection of moons, Saturn exceeds it in terms of the overall number of moons. These moons of Saturn are thought to have been formed through various processes, including the capture of smaller bodies and accretion from debris created by previous moon-forming impacts.

98. **S**aturn's Walnut-Shaped Moon. Pan is a celestial object orbiting Saturn, and it serves as a ring shepherd for the planet. However, what truly sets Pan apart is its unique shape, which has captivated many enthusiasts. Its irregular appearance resembles that of a walnut, which is a result of the accumulation of ring debris from the Encke gap.

99. **T**he Fascination of Saturn's Rings: Made of Millions of Ice Particles. Saturn's rings are not solid structures, as most of us may have thought, but are made of billions of ice particles ranging from tiny grains to chunks as large as houses. These particles continuously collide and break apart, resulting in a never-ending cycle of change. The gravitational influence of nearby moons also affects ring particles, resulting in intricate patterns and structures within the rings, such as the "spokes"

seen in the B ring, which are believed to be caused by the gravitational influence of the moon Prometheus. Saturn's rings are an unavoidable space symphony that cannot fail to astound.

100. **T**he **Flattened Rings of Saturn: The Flatness of its Mesmerizing Rings.** Saturn's flat rings are the product of a delicate balance of gravitational and other forces acting on water ice particles. While gravity pulls them towards the planet, other factors such as collisions with other particles and the solar wind maintain them in orbit. The size of the particles also influences flatness, with bigger particles forming a more stable and flat formation. Despite continuous changes, Saturn's gravitational pull and the balance of forces maintain the rings' flatness. The thickness of the rings varies considerably, with an estimated average thickness ranging from 10 to 100 meters, depending on where you are in the ring system.

101. **T**he Swift Dance of Saturn's Rings: Unveiling the Celestial Motion. Saturn's ring particles travel at breakneck speed due to the planet's gravitational force. The precise speed at which they travel is determined by their distance from Saturn, with particles closer to the planet moving faster than those further away. The ring particles move at speeds varying from 34,796.79 miles to 44,738.71 miles (56,000 to 72,000 kilometers) per hour on average. It is critical to remember that the velocity of the particles in the ring is variable and subject to change over time due to a variety of factors, such as particle interactions and the effect of solar wind. Despite these modifications, the ring's total speed remains constant.

102. **Saturn's Ring Rain: A Cosmic Downpour.** Saturn's rings are constantly changing. This process, known as "ring rain," refers to the constant loss of material from the rings, which is then replenished by ice particles from the planet's small moons. This ring rain causes the rings to slowly spiral towards the planet over time, making it appear as if Saturn is shedding tears. But this "rain" is not only visually stunning, but it also provides valuable information about the history of Saturn and the solar system. Scientists study this process to understand the dynamics of Saturn's ring system and the potential origins of its rings.

103. **Extraterrestrial Rain of Diamonds: Wealth in Jupiter and Saturn.** Saturn and Jupiter have been found to have large amounts of carbon in the form of diamond and graphite. This is due to the high pressures, and temperatures found deep within the planets' atmospheres. The carbon is thought to be present in the form of "diamond rain," where methane is compressed and heated to such an extent that it forms solid diamond particles that fall like rain. Similar diamond rain may also occur on Uranus and Neptune, with significant amounts of methane in their atmospheres.

"

*The most beautiful
experience we can have is
the mysterious. It is the
fundamental emotion that
stands at the cradle of true
art and true science.*

– Albert Einstein

SECTION 10
Uranus and Its Unconventional Characteristics

104. **U**ranus: **The First Planet Discovered.** Uranus was the first planet discovered beyond the five known since ancient times: Mercury, Venus, Mars, Jupiter, and Saturn. It was first observed by William Herschel in 1781 using a telescope.

105. **T**he **Mysterious Dark Rings of Uranus: Composition of the Planet's Unique Ring System.** Similar to Jupiter and Saturn, Uranus possesses a ring system that was only detected by Earth telescopes in 1977 due to the particles in the rings being dark gray. The multitude of narrow rings and dust bands are some of the darkest objects ever studied in our solar system, with most located within one planetary radius of the cloud tops. However, NASA's Hubble Space Telescope discovered a more distant pair of rings in 2003, and they are extremely faint. The narrow rings are primarily composed of large particles, with boulder- to house-sized fragments being dominant. The outermost Epsilon ring is composed mostly of black boulders, many of which are over 10 meters in diameter. The dust bands discovered by Voyager 2 also contribute to the ring system. Experts propose that the dark particles in the rings are carbon-rich residues or exotic black rocks containing organic compounds and ice. The rings lie in the equatorial plane of Uranus, suggesting that they were formed after Uranus fell on its side. If the rings existed before this event, it is unlikely that they would have reoriented their orbits following the impact.

106. **U**ranus' **Peculiar Axis and Uncommon Seasons: 98 Degrees of Tilt.** Uranus is known for its unusual tilt, with an axial tilt of 98 degrees, which means it essentially orbits the sun on its side. This means that its poles are where other planets have their equators, and its equator is where other planets have their poles. As a result, Uranus has extreme seasons, with one pole experiencing 21 years of darkness and the other 21 years of sunlight.

107. **U**ranus' **Unique and Weirdest Magnetic Field: A Tilted and Offset Maverick of the Solar System.** A Magnetic Maverick of the Solar System. With its magnetic field tilted at a remarkable 60 degrees and offset from its rotation axis, Uranus stands out as a one-of-a-kind planet in our solar system. This magnetic oddity is believed to stem from the planet's rotational dynamics, which generate its magnetic field in the atmosphere rather than in its core. This makes Uranus a subject of ongoing fascination for planetary scientists and astrophysicists, who are eager to learn more about its unique magnetic field and planetary dynamics. As Carl Sagan once said, 'Uranus is a never-before-explored planetary frontier.'

108. **U**ranus: **A Planet with a Dark Meteorite Impact Site Shaping its Unique Orientation.** Uranus is the only planet in the solar system that rotates on its side, and scientists believe that a massive impact with a dark meteorite caused this strange orientation. This impact is thought to have been so powerful that it knocked Uranus onto its side and changed the planet's rotation axis. The impact site on Uranus is believed to be a dark, crater-like feature visible in images taken by the Voyager 2 spacecraft. This dark impact site provides valuable clues about the early history of Uranus and the violent events that shaped the solar system.

109. **The Moons of Uranus: Dark and Frozen.** Unlike other planets in the solar system, Uranus' moons are all dark and frozen, with surface temperatures plummeting as low as -142 degrees Fahrenheit (-224 degrees Celsius). These moons are primarily composed of water ice and are relatively small compared to other moons in the solar system. The dark and frozen nature of Uranus' moons results from their distance from the sun and the lack of heat-producing internal activity.

110. **Shakespearean Skies: The Unique Naming of Uranus' Moons.** Uranus' moons are all named after characters from William Shakespeare's plays. The naming convention for Uranus' moons was established by its discoverer, Sir William Herschel, who chose to name the planet's first two moons after characters from Shakespeare's play "A Midsummer Night's Dream". Since then, additional Uranian moons have been named after characters from other Shakespearean plays such as "The Tempest," "Romeo and Juliet," and "Hamlet." This naming convention not only adds a literary touch to the study of Uranus' moons but also highlights the importance of artistic and cultural contributions to the field of astronomy.

111. **The Backward Spin of Venus and Uranus.** In contrast to the other planets within our solar system, Venus and Uranus possess an exceptional rotation pattern, wherein the sun appears to rise in the west and set in the east. Astronomers believe that Uranus collided with an Earth-sized object at some point, changing its rotation. At the same time, Venus orbits anti-clockwise due to an earlier event that knocked it off its upright position.

"

*Somewhere, something
incredible is waiting to be
known.*

– Sharon Begley

SECTION 11
The Blue Neptune and Planet 9

112. **T**he **Blue Beauty of Neptune: A Cosmic Mirage.** Neptune is famous for its brilliant blue color, which sets it apart from the other gas giants in our solar system. This blue hue is due to the presence of methane in Neptune's atmosphere, which absorbs red light and reflects blue light, creating the planet's distinct coloration. The exact process by which the methane in Neptune's atmosphere creates this blue color is still a subject of scientific investigation. Still, it is thought to be related to the planet's high atmospheric pressure and low temperatures. The blue color of Neptune adds to its otherworldly beauty and continues to be a source of fascination and intrigue for scientists and astronomers alike.

113. **N**eptune **Radiates More Heat than it Receives from the Sun.** The Sun does not provide the energy for such swift winds and ever-changing clouds. Neptune radiates 2.6 times more energy than it takes in from the Sun. Neptune appears to have an internal heat source, similar to Jupiter and Saturn, which causes it to radiate more energy than it takes from the Sun.

114. **N**eptune's **One Orbit: The Planet Discovered by Mathematics and Optical Observation.** Neptune, the outermost major planet in our solar system, has only completed one orbit around the sun since its discovery in 1846. Astronomers first discovered Neptune using mathematical predictions when they observed Uranus's deviant orbit. The planet's great distance from the sun makes it difficult to observe without optical equipment, and one year on Neptune is equivalent to 164.8 Earth years.

115. **Neptune Has the Fastest Winds in the Solar System.** Neptune's winds are considered the fastest in the solar system, reaching speeds of up to 1,500 miles per hour (2,400 kilometers per hour) driven by temperature differences between the equator and poles, creating large storms such as the Great Dark Spot, studied to understand the dynamics of the planet. These massive storms could swallow Earth whole, making them genuinely larger-than-life. One wonders if some otherworldly espresso fuels the planet's atmospheric gusts.

116. **The Great Dark Spot: A Stormy Mystery on Neptune.** The Great Dark Spot on Neptune is a massive storm system that was first observed by NASA's Voyager 2 spacecraft in 1989. This storm is roughly the size of the continental United States and appears as a dark, swirling vortex in the planet's atmosphere. While the exact cause of the Great Dark Spot is still a mystery, scientists believe it is likely formed by combining Neptune's internal heat with the planet's powerful atmospheric circulation. The Great Dark Spot has been extensively studied to understand the dynamics of the planet's atmosphere and is considered one of the most mysterious and awe-inspiring features in our solar system.

117. **Planet 9: The Search for the Elusive Giant.** In 2014, scientists proposed the existence of a hypothetical planet Nine, or VP113, or planet X, in the outermost regions of our solar system. This planet is thought to be between five and ten times more massive than Earth and has an elongated orbit about 20 times farther from the sun than Neptune. The observed patterns in the orbits of several small, distant objects in the Kuiper Belt provide compelling evidence for its existence. Its gravitational

influence is also believed to be responsible for the unusual clustering of distant Kuiper Belt objects in one particular direction. Researchers have presented mathematical proof supporting the potential existence of a ninth planet in our solar system, which they have dubbed "Planet Nine." This planet could be Neptune-sized and may have a mass approximately 10 times that of Earth, with an average orbit 20 times farther from the Sun than Neptune. It is believed to have a highly elliptical orbit that takes between 10,000 and 20,000 years to complete, and scientists are now searching for it using the world's most powerful telescopes.

Astronomy is a humbling and character-building experience.

– *Carl Sagan*

SECTION 12
The Curious World of Dwarf Planets

118. **D**warf Planets: The Outcast Celestial Objects of the Solar System. Dwarf planets exhibit a few traits, such as orbiting around a star, having a spherical shape, and having other large bodies close to them, such as asteroids, comets, and other dwarf planets. The fundamental difference between planets and dwarf planets lies in their size. Nine of the largest known candidates for dwarf planets, including Pluto, Eris, Haumea, Makemake, Gonggong, Quaoar, Sedna, Ceres, and Orcus, are generally agreed upon by astronomers to be classified as dwarf planets. Of these nine, two have been visited by spacecraft, and seven have at least one known moon, allowing for estimates of their masses and densities to be made. The masses and densities can then be used to determine the nature of these celestial objects. Only Sedna does not have any known moons and has not been visited, making it difficult to estimate its mass. There is no consensus among astronomers regarding the inclusion of smaller bodies as dwarf planets.

119. **P**luto Could Have Sustained Life in The Past. Some scientists have theorized that if Pluto had a liquid ocean and enough energy, it could potentially harbor life. However, the average surface temperature on Pluto is around -382 degrees Fahrenheit (minus 229 degrees Celsius), making it a frigid world unconducive to life as we know it. Additionally, recent analyses suggest that the probability of Pluto being habitable early in its existence is relatively low, based on our knowledge about the planet and its current conditions.

120. **P**luto's Dwarf Status: A Small World That Still Holds Big **Surprises.** Pluto, once considered the ninth planet in our solar system, is now classified as a dwarf planet. It is much smaller than Earth, with a diameter of only about 1,477.84 miles (2,377 kilometers), roughly one-fifth (0.2 times) Earth's size. Despite its diminutive size, Pluto is a fascinating world that has captured the interest of scientists and the public alike. It has a complex surface with features such as mountains, valleys, and plains and a thin atmosphere composed mainly of nitrogen, methane, and carbon monoxide. Its five known moons are also intriguing objects, each with unique characteristics.

121. **I**ce Mountain: Pluto's Surface. Pluto's diverse topography encompasses mountains, valleys, plains, and craters. With temperatures that can plummet to -375 to -400 degrees Fahrenheit (-226 to -240 degrees Celsius), some mountains reach elevations of 2 to 3 kilometers. The planet's icy mountains consist of massive blocks of water ice, occasionally coated with frozen gases like methane. Additionally, long troughs and valleys that stretch for up to 370 miles (600 kilometers) contribute to Pluto's intriguing features.

122. **P**luto's Icy Heart: Sputnik Planitia. Scientists have proposed multiple theories about how Pluto's heart-shaped feature, known as Sputnik Planitia, was formed. While many believe it was created by an impact basin, new research suggests that the nitrogen ice cap could have formed early on when Pluto was still spinning quickly. The theory proposes that the ice cap, once formed, would lock towards or away from Charon when Pluto's spin slowed to match the moon's orbital motion. Additionally, other models support the impact basin scenario and hint at the presence of a subsurface ocean on Pluto.

123. **E**ris: **The Dwarf Planet Beyond Neptune's Orbit.** Eris is categorized as a dwarf planet and is situated within the Kuiper Belt, which is a zone containing numerous frigid, icy objects positioned beyond the orbit of Neptune. Notable for being larger than Pluto, the once former ninth planet, Eris, was discovered in 2005 and played a crucial role in the reclassification of Pluto from a planet to a dwarf planet. Eris is a valuable object for astronomical study and exploration because of its unique properties. Studying Eris and other dwarf planets in the Kuiper Belt provides valuable insights into the formation and evolution of the solar system and the universe.

124. **A** Dwarf Planet with Water, Ice, and a Thin Atmosphere: **Ceres.** Ceres, located in the asteroid belt between Mars and Jupiter, is the region's largest object and is considered a dwarf planet. What sets Ceres apart is that it is believed to contain a substantial amount of water ice, which has led to speculation that it could be a future target for space missions seeking to extract water for use as a resource. Additionally, Ceres is distinctive for its bright spots and large impact crater, called Occator, and its weak atmosphere, making it one of only a few bodies in the solar system with a detectable atmosphere that is not a planet or moon.

125. **T**he Puzzling Bright Spots of Ceres: **The Enigma of the Dwarf Planet.** The dwarf planet Ceres, located in the asteroid belt, boasts a unique surface featuring bright white spots that have puzzled scientists for years. These spots, which are much brighter than the surrounding terrain, are believed to be deposits of salt and offer insight into the composition and history of Ceres and the asteroid belt as a whole.

126. **Haumea's Hula: The Surprising Discovery of Rings around a Dwarf Planet.** Haumea, a dwarf planet located in the outer reaches of our solar system, was thought to be a small, strange, irregularly elongated rock. However, in 2017, scientists discovered that Haumea has not one, but two rings made of ice particles. This makes Haumea the first known dwarf planet to have rings, and it raises questions about the formation and evolution of these rings.

127. **The Ice-Rich Rhea Montes: The Tallest Mountains on Dwarf Planet Ceres.** Rhea Montes, located on the dwarf planet Ceres, is a captivating feature in the asteroid belt between Mars and Jupiter. These mountains are not only some of the tallest features on Ceres, reaching a height of up to 2.5 miles (4 kilometers) above the surface, but they are also intriguingly composed of ice-rich material. This suggests that Ceres may contain a subsurface ocean beneath its icy crust, making the Rhea Montes an exciting subject of study for scientists and space enthusiasts alike.

128. **Exploring Makemake: A Dwarf Planet in the Kuiper Belt with a Reddish-Brownish Surface.** Makemake, one of the dwarf planets in the solar system, is located in the Kuiper Belt, a ring-shaped region beyond Neptune. Alongside Pluto, Eris, and Haumea, it completes one revolution around the Sun in about 305 Earth years. Makemake is slightly smaller than Pluto but is the second-brightest object in the Kuiper Belt when viewed from Earth. Its discovery alongside Eris has been significant in solar system studies, leading to the creation of the category of dwarf planets. Makemake's reddish-brownish color, similar to Pluto's, is visible from Earth. Scientists have detected frozen methane and ethane on its surface, with pellets of frozen methane potentially present on its cold surface. At its closest point to the Sun, Makemake may have a thin atmosphere made mostly of nitrogen.

> The moon is a loyal companion. It never leaves. It's always there, watching, steadfast, knowing us in our light and dark moments, changing forever just as we do. Every day it's a different version of itself. Sometimes weak and wan, sometimes strong and full of light. The moon understands what it means to be human. Uncertain. Alone. Cratered by imperfections.

— *Tahereh Mafi*

SECTION 13
The Moon: Exploring Earth's Celestial Companion

129. **The Moon: Born from Earth's Epic Collision.** The Moon's formation is believed to have occurred around 4.5 billion years ago due to a giant impact between the Earth and a Mars-sized object. The collision caused debris to scatter around the Earth, which later combined to form the Moon. Evidence of a shared geological composition between Earth and the Moon supports this theory, suggesting that the Moon is a fragment of the Earth created from material ejected during the impact event.

130. **The Moon's Proximity Shift: From Giant Moon to Celestial Companion.** In the past, the moon was located much closer to Earth, approximately 12-18,641 miles (20-30,000 kilometers) away. This close proximity would have resulted in the moon appearing 10 to 20 times larger in the sky, and its gravitational pull would have been much stronger. The tides would have also been intimidating eight times higher than they are today, making it difficult for any species to survive.

131. **The Moon's Vital Role in Sustaining Life on Earth: A Hypothesis.** The Moon plays a remarkable role in sustaining life on Earth, according to a hypothesis suggesting that life might not have evolved the way it did without it. The Moon's gravitational pull stabilizes Earth's rotation, regulating the planet's climate and protecting it from extreme changes. Additionally, the Moon's closer proximity to Earth in the past caused massive tides that led to fluctuations in salinity around coastlines, driving the evolution of early DNA-like

biomolecules. This hypothesis also suggests that life could not have begun on Mars. Overall, the Moon's impact on our planet and its environment has been critical to the creation and evolution of life on Earth.

132. **Confirmed: Water on the Moon - Discoveries by Lunar Mission Revealed.** Water on the Moon has been a topic of interest for scientists since the 1960s. The Lunar Reconnaissance Orbiter (LRO) has confirmed the existence of areas near the Moon's south pole where water is likely to exist. The LRO measured the amount of hydrogen in the lunar soil and found evidence of hydrogen-rich areas, indicating the presence of frozen water. Additionally, scientists have also measured the presence of water in the Cabeus crater on the Moon by crashing a spacecraft into it and analyzing the ejected chemicals.

133. **The Moon's Receding Act: Moving Away from Earth.** The moon and Earth have a unique relationship, as the moon is gradually receding from the Earth at the rate of 3.78cm (1.48 inches) per year, about the same speed at which our fingernails grow. The moon's gravitational pull draws certain parts of the Earth closer to it, causing the tidal bulge in our planet's oceans. Because the Earth rotates much faster than the moon, this bulge occurs slightly ahead of the latter, pushing the moon into a higher orbit around Earth due to various physical phenomena. The Moon may be drifting away from us, but it's still got us under its gravitational spell.

134. **The Lunar Illusion: The Moon's Asymmetrical Reality.** When the Moon appears to be full, it does indeed appear completely round to the naked eye. However, the Moon is not

perfectly symmetrical, and there are variations in its surface and interior. Some lunar basins contain significant masses beneath the surface that may be due to lava accumulations, and the Moon's crust is thicker on one side than the other. Scientists believe that this asymmetry and unevenness may have been caused by various factors, including an impact between the Moon and another object or the gravitational pull of the Earth.

135. **Lunar Eclipses: When Earth's Shadow Transforms the Moon's Appearance.** During a full moon phase, lunar eclipses can occur when the Earth is precisely positioned between the Moon and Sun, casting its shadow onto the lunar surface. This event can cause the Moon to dim and even take on a reddish hue over a period of a few hours. These eclipses can be viewed from half of the Earth. There are three types of lunar eclipses, including a total lunar eclipse where the Moon moves into the Earth's umbra and is dimly lit by sunlight passing through the Earth's atmosphere. This causes shorter wavelengths of light to scatter, making the Moon appear reddish or orangish. A partial lunar eclipse occurs when there is an imperfect alignment between the Sun, Earth, and Moon resulting in only a portion of the Moon passing through the umbra. Finally, a penumbral eclipse occurs during a full moon but is the faintest and most difficult to notice as the Moon passes through the outer part of Earth's shadow.

136. **The Moon Appear Red During a Lunar Eclipse. The Scientific Phenomenon of Rayleigh Scattering.** The Moon appears red during a lunar eclipse due to a scientific phenomenon called Rayleigh scattering. The same principle that makes the sky blue and sunsets red is responsible for the

Moon's red hue. Light moves in waves and different colors of light have varying physical properties. Blue light has a shorter wavelength and is more easily scattered by particles in Earth's atmosphere than red light, which has a longer wavelength. When the Sun is overhead, we observe blue light throughout the sky. However, when the Sun sets, sunlight must pass through more atmosphere and travel a greater distance before reaching our eyes. Blue light is scattered away, and longer-wavelength red, orange, and yellow light are transmitted. During a lunar eclipse, the only sunlight that reaches the Moon travels through Earth's atmosphere, causing the Moon to turn red. The more dust or clouds in Earth's atmosphere during the eclipse, the more intense the red color of the Moon. It appears as though the entire world's sunrises and sunsets are being projected onto the Moon.

137. **The Dark Side of the Moon: We See Only One Side of the Moon.** The Moon is in synchronous rotation with Earth, meaning it rotates on its axis at the same rate as it orbits our planet. This results in the same side of the Moon always facing Earth, and the far or "dark" side is hidden from view. The Moon's far side is a mystery, with a rugged and heavily cratered surface, a much thicker crust, and a different composition than the near side. Scientists believe the far side may hold vital clues to the Moon's early history and formation, making it a crucial target for future space missions.

138. **Moonquakes: Shaking Things Up on the Lunar Surface.** Earthquakes are known to be quick, sharp jolts, but have you ever heard of Moonquakes? These lunar trembles can last up to 10 minutes, 300 times longer than the average earthquake on Earth! According to NASA's Apollo missions, which placed

seismometers on the Moon's surface in the 1970s, these Moonquakes were caused by the gravitational tug-of-war between the Earth, the Moon, and the Sun, leading to the Moon's surface to be in a constant state of stress. Furthermore, Moonquakes can happen anywhere on the Moon, unlike earthquakes usually confined to plate boundaries.

139. **The Mystical Moonbow: A Rare and Captivating Rainbow Produced by Moonlight.** A moonbow is a natural phenomenon where a rainbow is produced by moonlight instead of direct sunlight. It occurs when light is refracted in water droplets, such as those in a rain shower or waterfall, and is always positioned opposite the Moon in the sky. Moonbows are fainter than solar rainbows, making it difficult for the human eye to see the colors, but they can be captured in long-exposure photographs. To view moonbows, the Moon must be at its brightest phase and low in the sky, not obscured by clouds, and the night sky must be very dark. Moonbows are rarer than rainbows produced by daytime sunlight and are most easily observed two to three hours before sunrise or two to three hours after sunset when the moon is rising or setting.

140. **The Moon's Slow but Steady Shrink: A Cooling Crust's Contraction.** Approximately 25% of the moon's seismic activity was caused by released energy from scarps or stairstep cliffs on the moon's surface. This activity is due to the contraction of the moon's crust as it cools, which has caused it to shrink by about 46 meters over the last hundred million years.

141. **The Ever-Changing Surface of the Moon: Constant Bombardment and Transformations.** The surface of the

moon is constantly evolving due to the high-energy particles and radiation it is bombarded with. This creates a layer of fine dust called regolith and features such as large rifts and valleys known as "rilles". The moon also has extensive, circular plains on its surface known as "maria," which were formed by ancient volcanic eruptions. The surface is further characterized by craters, mountains, and valleys and is mainly composed of rock.

142. **Tides: The Moon's Gravitational Pull on Earth's Oceans.** The Moon and Earth experience tides due to the Moon's gravitational pull on Earth's oceans. The Moon's gravitational pull is stronger on the side of Earth closest to it and weaker on the side furthest from it, causing the oceans to bulge and create high tides.

143. **Synchronous Rotation and Orbit: The Moon's Dance With the Earth and the Sun.** The Moon rotates in synchrony with its orbit around Earth, meaning the same side always faces Earth. Its rotation and orbit are influenced by the gravitational pull of both Earth and the Sun, causing its orbit to change slightly over time due to lunar perturbation and altering its rotation.

144. **The Moon's Surface Temperature: Hot and Cold.** The Moon's surface temperature varies greatly, from about -279.4 degrees Fahrenheit (-173 degrees Celsius) at night to 260.6 degrees Fahrenheit (127 degrees Celsius) during the day. This is due to the Moon's lack of an atmosphere to trap heat and its distance from the Sun. The temperature difference between the day and night side of the Moon can be as much as 752

degrees Fahrenheit (400 degrees Celsius). Such a remarkable contrast in temperature can cause thermal tension to build up on the surface, causing cracks and faults in the material.

145. **The Moon's Craters: A Remnant of Cosmic Collisions.** The Moon's surface is highly covered in countless sizes and shapes of impact craters. These craters are formed when asteroids, comets, and other celestial bodies collide with the Moon's surface. However, the Earth doesn't show the same shapes as in the past. This is because the Moon is not subjected to the same forces as the Earth. After all, it lacks water, an atmosphere, and tectonic activity. Due to this fact, the surface of the Moon hasn't experienced substantial alteration since its infancy, and most of its craters continue to be visible. Conversely, while over 80% of Earth's surface is less than 200 million years old, more than 99% of the Moon's surface is older than 3 billion years.

146. **Moon's Volcanic Activity: Fiery Past Shaped its Landscape.** The Moon is not currently volcanically active, but scientists have discovered evidence of past volcanic activity on its surface. This activity occurred billions of years ago, and the Moon's volcanoes were likely much larger and more powerful than Earth's. The Moon's volcanic activity helped shape its surface and created features such as the Moon's extensive plains known as maria.

147. **Moon's Heavy Metal: The Lunar South Pole's Titanium Deposit.** Recent findings from the Miniature Radio Frequency (Mini-RF) instrument imply that the Moon has elevated amounts of iron and titanium oxides exposed to the surface in larger craters between 3 and 12 miles (5 to 20 km) wide, which could explain the discrepancy in the current theory of how the

Moon was formed. Additionally, other research has shown that the Moon's South Pole has a significant deposit of titanium, along with high levels of ilmenite, which is also a significant source of lunar helium 3, a rare isotope that could be used as fuel in nuclear fusion reactors to produce clean energy. However, mining operations seeking to extract valuable resources may face challenges due to the extreme cold in the region, with some areas reaching as low as -396.4 degrees Fahrenheit (-238 degrees Celsius) due to being in permanent shadow.

148. **Moon-Earth Rock Exchange Program: Apollo 14's Lunar Collection.** Earth's oldest rock was found on the moon and brought home by Apollo 14. In 1971, Apollo 14 astronauts brought home various minerals and rock samples from their brief lunar trip. These samples were found to be around 4.5 billion years old, making them some of the oldest known materials in the solar system. So, technically speaking, the Moon holds a piece of Earth's ancient history. Millions of miles might separate the Moon and Earth, but the Apollo 14 mission has brought them closer than ever before. It's like a long-distance relationship that has defied the odds and succeeded, except instead of love letters, they exchange rocks.

149. **The Moon's Lunar Dust: A Persistent Dangerous Problem.** The Moon's surface is covered in fine, powdery dust called regolith, which is created by the constant bombardment of the Moon's surface by micrometeoroids and by the grinding of the Moon's surface by the action of the Moon's own soil. This lunar dust poses persistent challenges for lunar missions, as it can clog mechanical systems and damage equipment. Lunar dust is composed of sharp and abrasive particles that could

be harmful to humans. However, there is little information available about its toxicity. Despite this, the twelve individuals who have walked on the Moon have reported experiencing symptoms similar to hay fever, including sneezing and nasal congestion. Scientists are working on developing methods to mitigate the effects of lunar dust and protect equipment from its damaging effects.

150. **T**he Moon has an "Atmosphere": The Lunar Exosphere. The lunar body known as the Moon possesses a remarkably thin and rarefied atmosphere, which is referred to as the exosphere. Unlike the Earth's atmosphere, the Moon's atmosphere is not suitable for breathing due to its insubstantial nature. During the frigid lunar night, the exosphere descends to the ground. Helium, argon, sodium, and potassium are among the elements that make up this lunar atmosphere. In an exosphere, the gas particles are widely distributed, resulting in infrequent collisions with one another. They behave like tiny cannonballs, hurtling unimpeded across the curved, ballistic trajectories of the lunar terrain and bouncing off its surface. The atmosphere of the Moon contains only 100 molecules per cubic centimeter, a significant contrast to the 100 billion molecules per cubic centimeter in Earth's atmosphere at sea level. The combined mass of these gases on the lunar surface is approximately 55,000 pounds, equivalent to the weight of a loaded dump truck. The chilly nighttime temperatures cause the exosphere to drop to the ground, only to be re-stirred and raised by the solar wind in the days to come.

"

*The universe is not only
stranger than we imagine,
it is stranger than we can
imagine.*

– *Sir Arthur Eddington*

SECTION 14
Other Moons in The Solar System

151. **A** **Sky Full of Moons: Over 200 Natural Satellites in Our Solar System.** Our solar system is home to a vast number of celestial bodies, with over 200 moons orbiting various planets. While Jupiter and Saturn have the most significant number of moons, Mercury and Venus stand out as the only two planets without their own natural satellites.

152. **P**lanets **Without Moons in the Solar System: Mercury and Venus.** Mercury and Venus are the only planets in our Solar System with no natural satellites or moons orbiting around them. This peculiar feature is perplexing since all other planets, including Earth, have at least one moon. The reasons behind this lack of companionship for Mercury and Venus are still unknown and remain a topic of ongoing research in the field of planetary science. Several theories have been proposed, including the possibility that the gravitational pull of the Sun is too strong, preventing the formation of moons, or that collisions destroyed any moons that did exist with other objects. It has also been suggested that these two planets may have formed closer to the Sun than they do now, where the Sun's intense heat would have destroyed any moons or prevented their creation entirely. Whatever the cause, Mercury and Venus' lack of moons is a distinctive and fascinating feature that sets them apart from the other planets in the Solar System.

153. **C**allisto: **A World of Ice and Mystery with Potential for Life.** Callisto, the second-largest moon of Jupiter and the third-largest moon in our solar system, possesses a surface that bears the highest concentration of impact craters among celestial bodies in our solar system. Images taken by passing spacecraft have revealed bright white spots that contrast with darker regions on Callisto's surface. Scientists believe these bright areas primarily consist of ice, while the darker patches indicate eroded ice. Initially regarded as a geologically inert and lifeless rocky entity, data collected by the Galileo spacecraft in the 1990s challenged this notion by suggesting the existence of a subsurface ocean beneath Callisto's icy exterior. However, more recent studies have indicated that if an ocean exists, it might be situated at greater depths than previously hypothesized or potentially not exist. If indeed present, this ocean could possibly interact with the rocky material on Callisto, thereby creating a conceivable environment that could support life.

154. **The Extreme Orbital Distance of Nereid: A Rare Moon in Our Solar System.** Nereid is a moon of Neptune that is unique among the known moons in our solar system due to its extreme orbital distance. Unlike most moons, which orbit their host planets relatively close, Nereid orbits Neptune at a distance of more than 5 million kilometers. This places Nereid, after Triton, as the second most distant moon in our solar system.

155. **Rolling Rocks: The Chaotic Tumbling of Moons.** Several moons in our solar system, including Hyperion (a moon of Saturn) and Ananke (a moon of Jupiter), defy the norm with their chaotic rotations. Unlike most moons that maintain a steady spin, these celestial bodies tumble through space due to their irregular shapes and lack of a consistent gravitational pull. This fascinating phenomenon offers a glimpse into the unexpected diversity of the moon populations in our solar system. The chaotic tumbling of moons like Hyperion and Ananke is like the cosmic version of the jitterbug. They dance to their own tune, defying convention and reminding us that there is beauty in chaos.

156. **P**roteus: A Moon with Heavily Cratered Terrain. Proteus's surface is characterized by a patchwork of light and dark areas, with bright areas ten times brighter than dark areas. This high-contrast terrain is thought to be due to differences in the surface composition and reflectivity of the different landscapes on Proteus. In addition, the heavily cratered terrain suggests that Proteus is one of the oldest objects in the Neptune system, providing valuable insights into the early evolution of the Neptune system and the processes that shaped its moons. Proteus should be approached with utmost care lest it responds with a formidable barrage of craters.

157. **T**win Moon Phenomenon: Janus and Epimetheus. Janus, a potato-shaped moon, shares an orbit with Epimetheus in a co-orbital condition or 1:1 resonance, where they trade places between inner and outer orbits approximately every four Earth years due to their gravity interaction. They have a uniquely fascinating history, believed to have been a single moon that split in two after a catastrophic collision with another celestial body. Janus has several prominent craters while Epimetheus is more spherical. These two moons also share an atmosphere of charged particles, indicating they may be linked. Together, they trail enough particles to generate a faint ring, and their common orbit appears as a gap between Saturn's F and G rings. This is the first known case of a moon orbiting another moon, and it raises many intriguing questions.

158. **Iapetus: The Yin and Yang Moon.** Iapetus is a moon of Saturn that has a unique appearance, with one hemisphere being very dark and the other much brighter. Iapetus is believed to be made up of three-quarters ice and one-quarter rock and thermal segregation is responsible for its dark hemisphere. It also has an equatorial ridge made up of 6-mile-high mountains, with two theories on how it formed. Giovanni Cassini discovered Iapetus in 1671, and since then, scientists have been trying to determine why one hemisphere of the moon is so dark compared to the other and other surfaces in the Saturn system. Some theories suggest that Iapetus is sweeping up particles from the more distant dark moon, Phoebe, or that there might be ice volcanism distributing darker material to the surface. However, recent studies have shown that thermal segregation is likely the main cause of Iapetus' dark hemisphere. Additionally, Iapetus has an equatorial ridge made up of mountains, with two theories on how it formed: one suggests that it was formed at an earlier time when Iapetus rotated much faster, while the other proposes that it was made from material left from the collapse of a ring.

159. **A Unique Dance of Avoidance: Naiad and Thalassa.** Neptune, the eighth planet from the sun, boasts 14 known moons, the largest and most famous of which is Triton. Among Neptune's other moons are the small and irregularly shaped innermost satellites, Naiad and Thalassa, which are locked in a rare type of orbital resonance. This phenomenon results in Naiad orbiting around Neptune every seven hours, while Thalassa, positioned further out, takes seven and a half hours to complete its orbit. As Naiad gains four laps on Thalassa, an observer on Thalassa would see Naiad in a zigzag pattern,

passing by twice from above and twice from below. Although this dance may appear odd, it keeps the orbits stable. This unique choreography is believed to have originated from the breakup of a larger celestial body. While numerous celestial dances exist in our universe, this particular dance is unprecedented, making it a captivating subject for study and exploration.

160. **The Celestial Snowball: Dwarf Planet Ceres' Moon Vesta.** Vesta, a majestic celestial snowball and one of the most significant asteroids in our solar system, has a small moon discovered only recently. This unchanging rocky object has remained frozen since its formation, providing an invaluable glimpse into the solar system's earliest days, a precious time capsule preserving a unique record of our cosmic origins.

161. **Triton: The Retrograde Moon of Neptune.** Neptune's 13th moon, Triton, is a celestial wonder because it orbits its planet in reverse, making it the only moon in the solar system to do so. This unusual characteristic has intrigued scientists for years. Recent studies suggest that Triton was originally a Kuiper Belt Object that was eventually ensnared by Neptune's gravity millions of years ago. The captivating tale of Triton's capture and its subsequent retrograde orbit has added to its allure and continues to fascinate scientists.

162. **The Lunar Lighthouse: Neptune's Moon Triton. Like a beacon in the night sky, the moon Triton** illuminates Neptune with its dynamic and vibrant presence. An active world teeming with geysers, icy plumes, and a frozen nitrogen atmosphere, Triton harbors a subterranean ocean that holds the tantalizing possibility of extraterrestrial life. Truly one of the most enthralling celestial bodies in the solar system.

163. **The Largest Moon in The Solar System: Ganymede's Jumbo Size.** Ganymede is the largest moon in our solar system and one of Jupiter's loyal companions. With a diameter of 3,271 miles. (5,262 kilometers), it surpasses even the size of Mercury. Ganymede is composed of rock and ice and has a magnetic field, making it the only moon in the solar system to have one. It is believed that Ganymede may have a subsurface ocean of liquid water beneath its icy surface, making it a prime target for future exploration missions.

164. **Jupiter's Miniature World: Io, the Most Volcanically Active Moon.** Jupiter's smallest moon, Io, is a fiery, tempestuous world with volcanic activity unparalleled in the solar system. With over 400 active volcanoes, Io is a constantly evolving, geologically dynamic world shaped by Jupiter's and its neighboring Galilean moons' intense gravitational forces. Because of its volatile nature, it is a fascinating and captivating object of scientific investigation.

165. **Charon: The Massive Companion to Pluto.** Charon is a fascinating object for study. One of the most striking features of this moon is its enormous size compared to its host planet. As the biggest among Pluto's five satellites, it measures almost half of Pluto's own size, rendering it an exceptional example of the moon-to-planet size ratios observed in the entire solar system. Charon's size and proximity to Pluto have caused significant tidal forces, shaping the surfaces of both bodies and providing valuable information on the evolution of binary planet systems. Charon and Pluto maintain a captivating

bond that renders them the ultimate odd couple in our solar system.

166. **R**hea: **The Saturnian Moon with Grooved Terrain.** Rhea, a moon of Saturn, has a strange and grooved terrain that is thought to have been caused by tidal forces from Saturn. The cause of the grooves on Rhea's surface is believed to be the result of Saturn's gravitational forces, which stretched and reshaped the moon, leaving behind these distinctive features. Further study of Rhea could provide insight into the effects of tidal forces on moons and their evolution.

167. **T**he Martian Moons: **Two Captured Asteroids Orbiting Mars.** The two moons of Mars, Phobos, and Deimos, are believed to be captured asteroids based on their irregular shapes and heavily cratered surfaces. This evidence offers a distinctive viewpoint on the initial phases of the development of the solar system and the movements involved in the transformation of the planets. Phobos, the larger moon, also holds potential for scientific investigation and future exploration due to its proximity to Mars.

168. **P**hobos: **The Dusty Moon Spiraling Inwards to Mars.** Phobos, hollow and nearly shattered by a giant crater and battered by thousands of meteor impacts, is on its way to colliding with Mars. Phobos is the larger of Mars's two moons and is approximately 17 miles (27 kilometers) in diameter. It orbits Mars three times each day and is so close to the planet's surface that it is not always visible in some places on Mars. It is covered in a fine dust layer and slowly spirals inward towards the planet. Scientists believe this inward spiral will eventually lead to Phobos being torn apart and creating a ring system

around Mars. This process provides valuable information on the evolution and eventual fate of moons in the solar system.

169. **T**he Hidden Gems of Uranus: The Unknown Moons. Uranus, the seventh planet from the sun, has 27 known moons, many of which are relatively small and unknown. Among these, the most enigmatic and least studied are Cordelia and Ophelia. Cordelia, the innermost moon of Uranus, is only about 15 km in diameter and was discovered in 1986 by the Voyager 2 spacecraft. On the other hand, Ophelia is slightly larger at 12.4 miles (20 kilometers) in diameter and orbits Uranus at a distance of about 31,690 miles (51,000 kilometers). Despite their small size and distance from Earth, these moons continue to pique the interest of scientists and researchers with their peculiar characteristics and behaviors.

170. **T**he Patchwork Terrain of Miranda: A Heavily Cratered and Varied Surface. Miranda, one of the moons of Uranus, boasts a surface that is both unique and varied. Its heavily cratered terrain is interspersed with massive cliff formations and valleys that have high walls, likely created by tectonic processes and impact cratering. At the same time, Miranda's strikingly asymmetric profile, characterized by imposing ridges and valleys stretching over hundreds of kilometers and plunging several kilometers deep, is a result of the moon's early impact history. Thus, Miranda's patchwork terrain and unusual shape provide valuable clues into its early history and the evolution of the Uranian system.

171. **Active Oberon: The Erupting Moon of Uranus.** Among the 27 known Uranus moons, Oberon stands out as the most active. Oberon is the second largest of Uranus' moons, with a diameter of about 932 miles (1,500 kilometers). It was discovered in 1787 by William Herschel. Oberon is fascinating to scientists because of its geologic activity, including cryovolcanism and tectonic activity. Cryovolcanism is the eruption of water, methane, and ammonia, which create geysers and cryomagma. Tectonic activity is the movement of the surface of the moon, which creates mountains and valleys. Oberon is not the only active moon in Uranus; Ariel, Titania, and Umbriel also have some signs of geologic activity.

172. **T**he **Weirdest Moon in the Solar System: The Unusual Shape and Spin of Hyperion.** Hyperion, a Saturnian moon, is known for its unusual shape, surface characteristics, and erratic rotation. With a porous surface resembling a giant sponge or an irregular boulder, it lacks a stable rotation axis and instead tumbles unpredictably in space. This phenomenon is believed to be caused by the gravitational pull of Saturn and nearby moons, which destabilize Hyperion's rotation and cause it to chaotically rotate on its axis. Additionally, Hyperion's spin is tilted at 26 degrees, unlike most moons that orbit their planet with a rotation almost perfectly aligned with its equator. These peculiar characteristics make Hyperion one of the weirdest moons in the solar system, and while scientists still need to fully understand the dynamics that keep it in such a twisted position, it's probably due to its irregular shape and gravitational interactions with nearby moons. Despite its oddities, Hyperion's spongy appearance and lower density make it an object of fascination for planetary scientists studying the dynamics of celestial bodies.

173. **T**he **Unusual Composition of Titania.** Uranus' largest moon is thought to have a rocky silicate core surrounded by a layer

of water ice. This composition is unusual for a moon's size and suggests it may have formed differently than other moons in the solar system. Additionally, Titania is the only known Uranian moon with a dense atmosphere composed primarily of carbon dioxide. Scientists believe this atmosphere may result from outgassing from the moon's interior.

174. **Pluto's Five Intriguing Moons: From the Enigmatic Charon to the Reddish Nix and Hydra.** Pluto, the dwarf planet located in the Kuiper belt, has five known moons. The largest and most well-known of these is Charon, over half the size of Pluto itself and the largest known satellite relative to its parent planet. But Charon isn't the only moon worth noting. The moons Nix and Hydra were discovered in 2005 and are thought to be irregularly shaped and possibly elongated. They also have reddish regions, which scientists believe may be due to the presence of tholins, a type of organic compound. In addition to Nix and Hydra, Pluto has two smaller moons, Kerberos and Styx, which were discovered in 2011 and 2012, respectively.

"

We are all connected; To each other, biologically. To the earth, chemically. To the rest of the universe atomically.

– Neil deGrasse Tyson

SECTION 15
Space Rocks: Comets and Asteroids

175. **C**omets: **Fascinating Space Wanderers Made of Frozen Gases, Rocks, and Dust.** Comets are fascinating space wanderers that consist of frozen gases, rocks, and dust left over from the formation of the solar system. They orbit the sun in highly elliptical orbits and can take hundreds of thousands of years to complete a single orbit. As comets approach the sun, they heat up quickly, causing the solid ice to turn into gas through a process known as sublimation. This gas includes substances like water vapor, carbon monoxide, and carbon dioxide, and is eventually swept into a distinctive comet tail. Comets have a structure that includes a nucleus, coma, hydrogen envelope, dust, and plasma tails. They leave a trail of debris behind them that can lead to meteor showers on Earth. Scientists classify comets based on the duration of their orbits around the sun, with some comets being sun-grazers that smash right into the sun or get so close that they break up and evaporate. In antiquity, comets inspired both awe and alarm.

176. **K**uiper Belt Objects: **Icy Celestial Bodies That May Have Delivered Water to Earth.** Kuiper Belt Objects (KBOs) are a group of small icy celestial bodies primarily made up of rock and ice that orbit the sun beyond Neptune. They contain many small icy objects and are also known as the "third zone" of our solar system, being the birthplace of many comets. KBOs such as Pluto and Eris formed in the early stages of the solar system, and there are around 70,000 KBOs larger than 100 km and many smaller ones. Comets composed of rock, dust, ice, and other frozen chemicals may have played a crucial role in the origin of life on Earth by delivering water to our planet billions of years ago. Recent research has indicated that the water found in many comets may have a common origin with Earth's oceans, suggesting that many more comets than previously thought could have delivered water to Earth. New observations from NASA's Stratospheric Observatory for Infrared Astronomy (SOFIA) reveal that Comet Wirtanen, originating from the Kuiper Belt outside our Solar System, contains "ocean-like" water, highlighting a vast reservoir of Earth-like water in the outer reaches of the solar system that could have crucial importance for the development of life.

177.   **Cosmic Nomads: The Enigmatic Centaur Objects.** Centaur objects are small bodies that are similar to asteroids in size and comets in composition. They exist in the outer solar system, mainly between the orbits of Jupiter and Neptune, and are believed to have originated from the Kuiper belt. These objects, with a diameter of up to 155 miles (250 kilometers), travel in unstable orbits that cross the paths of the giant planets. They spend a short lifetime as Centaurs and are continually replenished from the Kuiper belt.

178.   **Asteroids with Ring Systems: Astonishing Discovery Around Chariklo.** In 2014, astronomers discovered rings around the asteroid Chariklo. This small asteroid located in the outer regions of our solar system was found to possess two narrow rings named Oiapoque and Chuí. The presence of rings around Chariklo was unexpected, as they are typically associated with larger bodies like planets. This discovery indicates that even smaller celestial objects like asteroids can host ring systems. Observations during stellar occultations provided evidence of the rings' existence. Studying these ring

systems contributes to our understanding of the formation and dynamics of celestial bodies in our solar system.

179. **Halley's Comet: The Celebrity of the Sky.** Arguably the most well-known comet in history is Halley's Comet. As a "periodic" comet, it returns to Earth's vicinity about every 75 years, making it possible for a person to see it twice in their lifetime. It was last seen in 1986, and its next anticipated return date is 2061.

180. **The Solar Trans-Neptunian Objects: The Cosmic Explorers.** These are small planetary bodies that orbit the Sun at distances beyond Neptune, and there are currently over 3,000 known TNOs with possibly many more. The term "trans" in Latin refers to "beyond," and "trans-Neptunian" means beyond Neptune. As such, TNOs are primarily located in the Kuiper Belt, which lies beyond Neptune's orbit.

181. **Extragalactic Objects: Beyond the Milky Way.** Extragalactic objects are celestial bodies that exist outside of our Milky Way galaxy. These objects may consist of galaxy clusters, other galaxies, and more. Because of their distance, extragalactic objects are often challenging to observe and study. Nevertheless, studying these objects can provide crucial insights into the evolution and formation of galaxies, the structure of the universe on a large scale, and the nature of dark energy and matter. Astronomical tools, like the Hubble telescope, have shown that the universe is comprised of 100-200 billion galaxies, 500 billion planets, and 200- 400 billion stars. Given the vastness of these numbers, it is likely that there is more than one intelligent life form outside of planet Earth, according to many scientists.

182. **A**steroids vs. Comets: The Battle of Composition and Formation in the Early Solar System. When discussing the differences between asteroids and comets, their composition is the primary factor. Asteroids are primarily composed of metals and rock-like materials, whereas comets are made up of a combination of ice, dust, and rock-like substances. Both asteroids and comets are remnants of the early solar system, dating back 4.5 billion years. However, their formation processes and locations differ. Asteroids formed much closer to the Sun, in a region where the high temperatures prevented ice from staying solid. On the other hand, comets formed further away from the Sun, where ice could remain frozen. As comets approach the Sun, they begin to lose materials with each orbit, as the heat melts and vaporizes some of their ice, producing a visible tail. Interestingly, asteroids are believed to be remnants of the early solar system that never formed into full-fledged planets due to the massive gravitational influence of Jupiter disrupting their formation.

183. **C**omets from the Oort Cloud and Kuiper Belt: Comets May have played a role in the origin of life. Comets, composed of rock, dust, ice, and other frozen chemicals, may have played a crucial role in the origin of life on Earth by delivering water to our planet billions of years ago. Recent research has indicated that the water found in many comets may have a common origin with Earth's oceans. Until now, all the comets that have been measured have been believed to have formed early in the history of the Solar System in the region of the giant planets Neptune and Uranus and kicked out to a great distance as they bumped into the planets and each other. However, new observations from NASA's Stratospheric Observatory for Infrared Astronomy (SOFIA) reveal that Comet Wirtanen, originating from the Kuiper Belt outside our Solar System, contains "ocean-like" water, suggesting that many more comets than previously thought could have delivered water to Earth. The study highlights a vast reservoir of Earth-like water in the outer reaches of the solar system that could have crucial importance for the development of life.

184.   **The Potential for Asteroid Impacts: A Low but Real Threat.** Asteroids are rocky objects that orbit the sun and can pose a threat to Earth if they collide with our planet. A 100-kilometer asteroid impact could cause mass extinction, but the probability of such an event happening in the next billion years is low, nevertheless not zero. Crater counting is a method used to assess the probability of impact, and an equation proposed by Hartmann determines the probability of impact based on the diameter of the crater. Extrapolations from Ivanov's model suggest a low probability of a 1000-kilometer-wide crater event in the next billion years. However, such models should be taken with caution due to the limited amount of data available.

185.   **The Threat Posed by Near-Earth Asteroids and Comets: Potential for Devastation and Destruction.** Asteroids and comets that orbit near the Sun pose a threat to Earth if they come within 4.6 million miles and are at least 460 feet in diameter. If an object of this size were to impact Earth, it could cause regional devastation and destroy an entire city. Even larger objects could cause global effects and mass extinctions. Smaller objects can also cause significant damage, as seen in the Tunguska event in 1908 when an explosion over Siberia leveled more than 80 million trees. In 2013, an asteroid only 65 feet across caused significant damage and injured over 1,100 people in Russia.

186. **The Tunguska Event: The Comet or Asteroid That Impacted Central Siberia.** In 1908, an enormous explosion occurred in central Siberia near the Podkamennaya Tunguska River, flattening 2,000 square kilometers of land, and charring over 100 square km of pine forest. The explosion, estimated to have occurred at an altitude of 5-10 kilometers, is thought to have been caused by a comet or asteroid. The energy released from the explosion is estimated to be equivalent to as much as 15 megatons of TNT, making it a thousand times more powerful than the atomic bomb dropped on Hiroshima. Scientists have found small fragments of the object, but no impact crater was created due to the explosion occurring high in the atmosphere. The blast wave ignited the forests, but the subsequent wave extinguished them, leaving behind only charred remains.

187. **The Impact of Asteroids and Comets on Earth and the Solar System.** The solar system's craters are caused by impacts from asteroids and comets. While Earth's atmosphere protects it from most small debris, bigger fragments can cause property damage and injury. Larger fragments of 10 meters or more can cause considerable damage. Impacts of this size occur once per decade and those of the 100-meter class occur once every 1,000 years. There are over 100 impact craters on Earth, and the Ries Crater in Bavaria was formed by an asteroid or comet 15 million years ago. Chicxulub is the largest known impact crater, and geologists are unsure about several larger ring-like structures. There are over 150 asteroids known that come closer to the Sun than the outermost point of Earth's orbit.

188. **The Most Famous Collision of an Asteroid in Earth's History: The End of the Dinosaurs.** Around 66 million years ago, the Cretaceous-Paleogene extinction event wiped out about 80% of life on Earth, including all dinosaurs except for birds. The asteroid impact theory is the most widely accepted explanation for the mass extinction, which suggests that a huge asteroid struck Earth near the Yucatan Peninsula in Mexico and created the Chicxulub crater, which is about 180 kilometers (112 miles) across and 2 kilometers (1.24 miles) deep. The impact caused tsunamis, earthquakes, and landslides worldwide, spewing 10,000 billion tons of carbon dioxide into the atmosphere and altering the climate of the entire planet. The extinction of plant life, caused by the smoke blocking the sun, led to the demise of herbivorous dinosaurs and, in turn, the carnivorous dinosaurs that preyed on them.

189. **The Oort Cloud: The Edge of the Solar System.** The Oort Cloud is a vast and round collection of small, icy bodies that orbit around the Sun at distances far beyond Neptune, extending from about 2,000 to 200,000 astronomical units (AU) away from the Sun. The cloud is believed to contain trillions of objects, each less than 100 kilometers in diameter and weighing a total of 10- 100 times the mass of Earth. Although it cannot be seen directly, it is thought to be the origin of most long-period comets. The Kuiper Belt is another source for short-period comets. The Estonian astronomer Ernest J. Öpik suggested in 1932 that there may exist a distant supply of comets to replace those that burn out on their journeys through the inner solar system. Jan Oort computed the original orbits of 19 comets in 1950 and showed that 10 of them were fresh and came from the same distant location, leading to the discovery of the Oort Cloud.

190. **The Asteroid Belt: The Solar System's Rocky Ring.** Situated between Mars and Jupiter's orbits, the asteroid belt is an assemblage of millions of small rocky bodies, which are believed to be leftover fragments from the formation of the solar system. These asteroids range in size from a few meters to several hundred kilometers and encompass intriguing objects like the dwarf planet Ceres and asteroid Vesta. Nevertheless, due to the vast expanse of the asteroid belt, a considerable fraction of it remains unexplored and unknown.

191. **Beyond Asteroids: The Fascinating World of Space Rocks and Their Impact on Earth.** The asteroid belt contains millions of asteroids that range in size from a few meters to several hundred kilometers. It is situated between the orbits of Mars and Jupiter and is believed to be composed of leftover fragments from the formation of the solar system. Besides asteroids, the asteroid belt also contains other rocky objects in space that are smaller than planets and are classified as either comets, asteroids, or meteoroids. When meteoroids enter Earth's atmosphere and streak across the sky, they are called meteors or shooting stars. If they survive the fiery descent and land on Earth, they are referred to as meteorites.

192. **Trojan Asteroids: The Loyal Followers of Planets.** These celestial bodies share an orbit with a planet or a moon and remain at a stable point ahead or behind it in a gravitational dance that can last millions of years. Jupiter is the king of Trojans, with thousands of known objects in its orbit. These asteroids are believed to be remnants of the early solar system and may provide prominent clues to its formation and evolution. Trojan asteroids offer a unique opportunity to study

the building blocks of our solar system, and some scientists even speculate they could be potential targets for future space missions.

193. **M**oons Aren't Just for Planets: Over 400 Asteroids in the Solar System Have Moons. Moons can orbit not only planets but also asteroids in the Solar System. In fact, over 200 asteroids with moons have been detected in the Main Belt alone, and the total number exceeds 400 if trans-Neptunian objects are included. Asteroids with moons are formed when one asteroid collides with another at a specific speed and angle, causing a chunk of rock to be ripped off and sent into orbit. Sometimes, a collision can split an asteroid into two similarly sized parts that orbit the Sun together. The first confirmed discovery of a moon orbiting an asteroid occurred in 1993 during the Galileo mission when an even smaller asteroid named Dactyl was found orbiting asteroid 243 Ida. This discovery has helped scientists learn more about asteroids and the formation of moons.

194. **A**n Asteroids with its Own Moon: Dactyl's Case. Dactyl, also known as 243 Ida, is a small moon that orbits the asteroid 243 Ida within our solar system. It was the first asteroid moon to be discovered, unlocking a new level of understanding about the mysteries of our universe. Dactyl, measuring approximately 0.93 miles (1.5 kilometers) in diameter, is thought to have formed from a collision between 243 Ida and another celestial object. The study of Dactyl offers valuable insights into the complex interplay of forces that shape our solar system, shedding light on the existence of small moon systems within the asteroid belt.

195. **S**ome Asteroids Contain Valuable Minerals That Could Be Mined in the Future. Asteroids, particularly two metallic ones orbiting Earth, could be valuable sources of precious metals worth an estimated $11.65 trillion. These asteroids contain more iron, nickel, and cobalt than all the metal reserves on Earth combined and are considered possible targets for future asteroid mining. They could provide cost-effective metals for space exploration and help extend humanity's reach by exploring space. Smaller asteroids also offer insights into their larger counterparts, which could aid in assessing their potential for mining. Additionally, researchers propose that microbial-based space technology could be utilized to produce energy, mine resources, and create pharmaceuticals, leading to a sustainable future in space exploration and utilization. Overall, asteroid mining and space technology hold immense potential for the future of space exploration and resource exploitation.

196. **D**ifferences between Meteoroids, Meteors, and Meteorites. All three terms—meteor, meteoroid, and meteorite—are associated with the luminous trails we see in the sky. The name we use depends on where the object is located. Meteoroids refer to rocks or particles in space, ranging in size from dust to small asteroids. They are commonly known as "space rocks". When a meteoroid enters Earth's atmosphere or that of another planet, it creates a fiery streak known as a meteor or "shooting star". On the other hand, if the meteoroid manages to survive the fiery descent through the atmosphere and lands on the ground, it is called a meteorite.

197. **M**eteor **Showers Occur: The Science Behind Annual Spectacles.** Annually or at set intervals, the Earth moves through a trail of dusty debris left by a comet, resulting in meteor showers. The meteors are commonly named after stars or constellations near the area where they appear in the sky. The Perseids meteor shower, which occurs in August every year, is one of the most well-known examples of this phenomenon.

198. **D**idymos **and Dimorphos: Binary Asteroid System and Planetary Defense.** A Binary Asteroid System with Potential for Deflection. Didymos, also known as 65803 Didymos, is a unique and interesting celestial body in our solar system. It consists of two separate objects, Didymos A and its moonlet Dimorphos, orbiting each other. Didymos A, the larger of the two, has a diameter of approximately 780 meters, whereas the smaller moonlet Dimorphos has a diameter of approximately 160 meters. The study of Didymos and Dimorphos presented an exciting opportunity for scientists to study the properties and dynamics of binary asteroids and explore ways to deflect or redirect them if they pose a threat to Earth.

199. **H**istoric **Moment: NASA's DART Mission Successfully Deflects Asteroid Moonlet.** On September 26, 2022, NASA's Double Asteroid Redirection Test (DART) spacecraft successfully collided with the moonlet a. This marked the first time in history that humans have intentionally altered the motion of a celestial object and demonstrated the effectiveness of asteroid deflection technology. The impact happened 6 million miles (11.7 million kilometers) from Earth. On October 11, 2022, NASA announced that the mission had successfully altered the moonlet's orbit. Although the celestial bodies Dimorphos and Didymos, which are a pair of twins, were never considered a danger to the human population, NASA provided that it was possible to deflect a potentially hazardous asteroid that could collide with Earth.

200. **Dynamic Asteroid Didymos Ejecting Rubble into Space: ESA's Hera Mission Observations.** Recent findings suggest that the asteroid Didymos is rotating so rapidly that it is flinging material off its surface, potentially observable by the European Space Agency's Hera mission. This discovery shows that asteroids are not just stationary objects but can be dynamic and have significant effects on their surrounding space environment. The research was conducted using remote observations and computer modeling since direct data cannot be collected from Didymos from Earth. By analyzing the asteroid's shape, size, mass, and composition, the team of researchers determined that Didymos's quick rotation results in a bulge at its equator, causing material to accelerate outward with more force than the asteroid's gravity can counteract. The study also proposes a theory on how Didymos' companion, Dimorphos, may have been formed through the shedding of material from Didymos. However, this theory is still based on remote observations and has yet to be confirmed. Overall, this study provides valuable insights into the dynamic nature of asteroids and raises the possibility that other asteroids across the solar system may be ejecting material into space, highlighting the need for further research into these fascinating celestial objects.

201. **The Cosmic Outcast: The Strange Characteristics of the Asteroid Elst Pizarro.** Elst Pizarro is an asteroid that has a highly elongated shape and a unique orbit. It is also one of

the closest asteroids to Earth, coming within 7 million miles (11,265,380 kilometers) of our planet. The cause of Elst Pizarro's unusual shape and orbit is not yet understood, but scientists believe it may be the result of a past collision with another celestial body.

202. **Oumuamua: A Mysterious Visitor from Another Star System.** Oumuamua is a remarkable interstellar object, the first of its kind to have visited our solar system, and was discovered by the University of Hawaii's Pan-STARRS1 telescope on October 19, 2017, with funding from NASA's NEOO program. Initially classified as a comet, it was later reclassified as an asteroid until new measurements revealed its slight acceleration, indicating it behaves more like a comet. This rocky, cigar-shaped object with a reddish hue is up to a quarter mile (400 meters) long and has an unprecedentedly elongated shape with an aspect ratio of perhaps 10 times as long as it is wide. It is believed to have wandered unattached to any star system for hundreds of millions of years before its chance encounters with our solar system. Although observations of 'Oumuamua vary in brightness by a factor of 10 as it spins on its axis every 7.3 hours, it is completely inert without any hint of dust around it, composed of rock and possibly metals, and has no water or ice on its surface. Its outbound path is about 20 degrees above the plane of planets that orbit the Sun, and it will head for the constellation Pegasus as it leaves our solar system.

203. **Arrokoth: A Fascinating Interstellar Asteroid That Preceded Oumuamua.** In 2014, astronomers observed a small space rock entering Earth's atmosphere, suspected to have originated outside our solar system due to its trajectory.

Further observations revealed that the 2014 MU69 or Arrokoth object, is a rare type of asteroid originating from another star system. This groundbreaking discovery was made in 2018, before the famous interstellar visitor 'Oumuamua, and has provided valuable insights into the formation and evolution of planetary systems beyond our own.

,,

*We are a way for the cosmos
to know itself.*

– Carl Sagan

SECTION 16
Time and Space: Journeys, Theories, and Challenges

204. The Possibility of Time Travel: Einstein's Theory of Relativity and Its Real-Life Applications. Technically, we all travel in time, but when we think of time travel, we usually imagine traveling faster than 1 second per second. However, science says it's possible, as Einstein's theory of relativity shows that time and space are linked together, and the faster you travel, the slower you experience time. This has been proven in experiments, such as the one with the airplane and the clocks. While we can't time travel hundreds of years into the past or future like in movies, we do use time travel calculations in everyday life, such as in GPS satellites. Due to the satellites' orbit and Earth's weaker gravity at that altitude, the clocks on GPS satellites experience time at a slightly faster rate than clocks on the ground, but scientists correct for these differences in time to ensure GPS accuracy.

205. The Time-Travelling Light: From Sun's Core to Earth's Sky. The journey of light from the core of the Sun to its surface takes a staggering 170,000 years, despite the light from the surface reaching us quickly. This time-traveling light serves as a reminder of the immense power that we take for granted.

206. A Journey Through Time: Gazing into the Sky Unveils the Universe's Past. When we gaze into the sky and see the light from stars and galaxies, we are looking back in time. This is because light takes millions or even billions of years to reach us, and it reveals the universe's past. Understanding this concept, called "light travel time," is crucial for comprehending the universe.

207. **A** **Journey to the Red Planet: Travel Time to Mars.** It takes around 6 to 9 months to travel from Earth to Mars, and another 3–4 months to travel from Mars back to Earth. The total trip duration is around 21 months, and it requires proper positioning of Earth and Mars, which only happens every 26 months. The optimal Mars launch windows for this decade are late 2023, late 2024, late 2026, and late 2028 or early 2029. During the journey, the spacecraft travels at a speed of around 24,600 miles per hour (39,600 kilometers per hour) for seven months and covers 300 million miles (480 million kilometers) to reach Mars. To explore the surface of Mars, fuel must be burned to get out of the elliptical orbit around the Sun and into Mars orbit, and additional fuel is needed to launch the lander off the planet's surface. Part of the spacecraft remains in Mars' orbit with supplies for the trip home while the crew explores the surface. Due to the trip's long duration, a lot of supplies, including food, water, clothes, medical supplies, and radiation shielding, are needed. A crew of six would require an estimated 3 million pounds of supplies. Being in space for a prolonged period can have physiological consequences, such as muscle atrophy and a weaker heart.

208. **C**hallenges of the Journey to Jupiter: Gas Giants, Crushing **Pressures, and Extreme Conditions**. Jupiter is a gas giant that is challenging to study due to its gaseous composition and extreme environment. The planet's atmosphere lacks oxygen, is incredibly hot, and has powerful winds due to its fast rotation. Landing on Jupiter is impossible because it lacks a stable surface, has crushing pressures, and extreme temperatures and pressures at deeper levels. The Galileo probe sent by NASA only lasted 58 minutes before losing contact due to the crushing pressures. Communication becomes impossible due to Jupiter's deep atmosphere absorbing radio waves. At around 430 miles down, the pressure and temperature become too high for any spacecraft to endure. At 13,000 miles down, temperatures are hotter than the surface of the sun, and the pressure is 2 million times stronger than on Earth, resulting in the formation of metallic hydrogen. Despite these challenges, scientists can still study and admire this mysterious planet from a distance. Traveling from Earth to Jupiter takes around six years.

209. **The Transformative Overview Effect: A Journey to a New Perspective**. The Overview Effect is a profound phenomenon that astronauts experience during spaceflight. By seeing the planet from space, they gain a new perspective on Earth and humanity that leads to a deep sense of awe, amazement, and interconnectedness with all life on Earth. This experience heightens their awareness of the fragility of our planet and the interdependence of all its systems. The transformative nature of the Overview Effect changes the way astronauts see the world and their place in it, leading to a greater appreciation for the need to protect the environment and promote global cooperation. In summary, the Overview Effect offers a unique and life-changing perspective that inspires a new way of thinking about our planet and our role in it.

210. **Theorized Journey Through Space and Time: The Enigmatic Wormholes.** Wormholes, the theoretical passages through space-time that could potentially offer shortcuts for long journeys across the universe, were first proposed in 1916 and elaborated on in 1935 by Albert Einstein and Nathan Rosen. They are expected by the theory of general relativity but have not been discovered to date. Despite their potential benefits,

wormholes also carry risks such as sudden collapse, radiation, and exposure to exotic matter. While black holes can form the mouths of wormholes, according to some solutions to general relativity, naturally occurring black holes alone cannot create a wormhole. It has been suggested that the presence of exotic matter or dark matter could sustain a wormhole, but these ideas remain theoretical and require more observational evidence. Wormholes are a fascinating concept in physics, but their existence has yet to be confirmed, and ongoing research continues to explore their possibility.

"

We do not realize that we are a part of the universe; we are the universe.

– Henry Miller

SECTION 17
The Fascinating Realm of Black Holes

211. **M**ilky **Way Could Harbor Up to 100 Million Black Holes in our Milky Way Galaxy.** Astronomers estimate that there could be up to 100 million black holes in our Milky Way galaxy. For a long time, scientists have not been able to confirm the existence of a solitary black hole. However, after six years of careful observations, the Hubble Space Telescope has finally provided concrete evidence for a lone black hole moving through interstellar space by accurately measuring its mass. Previous estimates of black hole masses have been made through statistical inference or by observing their interactions with other celestial bodies. Typically, black holes are discovered alongside companion stars, making this isolated black hole a rare and unusual find.

212. **T**he First Suspect Black Sheep: The Discovery of Cygnus **X-1.** Cygnus X-1 is the first object identified as a black hole. It was discovered in 1964 by astronomers at the Massachusetts Institute of Technology (MIT) and Harvard College Observatory. The discovery was made through observations of the X-ray emissions coming from the Cygnus constellation. The intense X-ray emissions and other observations provided strong evidence that a black hole was present in the system. In a casual bet made in 1974 between the scientists Stephen Hawking and Kip Thorne, Hawking bet that the Cygnus X-1 was not a black hole; however, he later admitted that he lost the bet, which was considered the first experimental evidence for the existence of black holes.

213. **T**he Birth of a Black Hole: From Star Collapse to Singularity. The formation of black holes begins with the collapse of a massive star. When a massive star runs out of fuel, its core begins to contract, becoming increasingly dense, and its gravity becomes stronger and stronger. When the core becomes dense enough, its gravity becomes so strong, that it eventually collapses under its own gravity that not even light can escape, resulting in fascinating phenomena such as black holes and spacetime singularities. These singularities, where physical quantities take arbitrarily large values, can either be covered within an event horizon or be visible to observers far away in the universe.

214. **F**our Ways Black Holes Form: From Stellar Collapse to LHC Speculations. We currently know four ways that black holes can form. The most understood is stellar collapse, where a large star runs out of nuclear fusion and the matter falls towards its own gravitational center, becoming so dense that a black hole is formed. These are known as solar mass black holes. Supermassive black holes, found in the centers of galaxies, are believed to have grown from smaller black holes by consuming other stars, but the details are still unclear. Primordial black holes could have formed from early universe density fluctuations, and there is a speculative idea that tiny black holes could form at the LHC if the universe has additional dimensions.

215. **S**ingularity at the Heart of Black Holes: Implications for Quantum Gravity. Black holes are regions of spacetime that contain a singularity, where the laws of physics break down. Our current understanding of spacetime, known as general relativity, suggests that singularities are inevitable in some real-world situations. However, black holes' singularity creates additional issues and queries, especially when considering quantum effects. Black holes appear to be thermodynamic entities when viewed through a quantum lens, raising the question of how our three fundamental theories relate. The development of black holes seems to contradict the conventional quantum evolution, sparking a debate on what physical laws a full quantum theory of gravity should uphold. These problems require careful philosophical consideration.

216. **W**here Time and Space Cease to Exist: The Fascinating Phenomenon of Time Dilation in Black Holes. One intriguing aspect of black holes is the ability to warp time through time dilation. As objects approach the black hole's center, time slows down until it appears to stop from an outside observer's perspective. Time dilation has been scientifically validated via several observations and has significant implications for our understanding of time and its link to space. When considering quantum effects, black holes seem to become thermodynamic entities, pointing to a connection between our three fundamental theories. However, the evolution of black holes seems to be in conflict with the standard quantum evolution, resulting in a discussion about the physical laws that a full quantum theory of gravity should maintain. These problems necessitate philosophical inquiry.

217. **T**he Information Loss Paradox: Quantum Entanglement Offers New Insights into Black Holes. Black holes emit Hawking radiation, causing them to lose mass over time. However, this raised an issue in physics, known as the information loss paradox, as it appeared to violate the principle

of information conservation, where information cannot be destroyed. Physicists believe that the solution to this paradox lies in the notion that Hawking radiation contains information. In 2019, new research provided a potential resolution by understanding spacetime and quantum entanglement. Researchers theorize that part of the interior of a black hole, called an "island," may be secretly on the outside due to quantum entanglement. This finding may offer new insights into the fundamental laws of the universe and how they operate within the extreme conditions of black holes.

218. **Runaway Star Rips Apart Near Black Hole, Causing Disappearance of X-ray Light**. X-ray light from black hole coronas, which are collections of ultrahot gas generated by the black hole's feeding, can exhibit noticeable changes in their luminosity. The sudden disappearance of X-ray light from a black hole corona in a galaxy known as 1ES 1927+654 faded by a factor of 10,000 in just 40 days. According to a new study, a runaway star might have come too close to the black hole and been torn apart, causing fast-moving debris from the star to crash through part of the disk, briefly dispersing the gas that feeds the black hole corona and causing the X-ray light to disappear. This theory could shed light on why the X-ray light from black hole coronas can exhibit noticeable changes in their luminosity, and it highlights the ongoing efforts of astronomers to better understand the mysterious behavior of black holes and the effects they have on their surroundings.

219. **Hawking Radiation: A Quantum Challenge to Classical Black Hole Theory.** Hawking radiation is a phenomenon predicted by Stephen Hawking in 1974-75, which states that a black hole should emit thermal radiation due to quantum

effects, in contradiction with the classical view of black holes as perfect sinks of energy and matter. This prediction has been supported by various mathematical derivations and physical models based on quantum field theory on curved spacetime. However, the foundational interpretation and physical origin of Hawking radiation remain contentious due to the various assumptions and approximations involved in these models, as well as the absence of a distinguished vacuum state or particle in relativistic spacetime. Nevertheless, this phenomenon provides a unique opportunity to study quantum field theory in extreme gravitational conditions and may lead to new insights into the nature of spacetime and quantum gravity.

220. **The Devouring Trap: Light Can't Escape Black Holes.** Black holes have a gravitational pull so strong that even light can't escape, which makes them a devouring trap. A black hole is a region where spacetime is highly curved, that every possible path for light eventually curves and twists the magnetic field back inside the black hole. Therefore, once a ray of light enters a black hole, it can never exit.

221. **Singularities: The Infinite Curvature of Spacetime and Its Possible Existence.** Singularities are points in spacetime where the curvature becomes infinite and general relativity breaks down. According to the singularity, theorems proved in the late 1960s, our universe began with an initial singularity, the Big Bang, approximately 14 billion years ago, and will likely end in a singularity as well. However, there is no commonly accepted definition of singularities. Some argue that singularities are too repugnant to be real, while others accept general relativity's prediction of singularities as consistent with the possible features of our world. The article examines

the different viewpoints on the need for a precise definition of singularity, and whether there is a need for one at all. The question of the existence of singular structures bears on the larger question of the existence of spacetime points in general.

222. **T**he **Existence of Massive Black Hole Triples.** Massive black hole triples are systems of three massive black holes that are gravitationally bound and in close proximity to one another. Observations of massive black hole triples provide valuable information about the evolution and growth of these oversized objects, as well as the formation and evolution of galaxies. They also offer the possibility of studying strong gravitational interactions and testing general relativity in extreme environments.

223. **A** **Supermassive Black Holes at The Center of Every Galaxy.** Most galaxies, including our own Milky Way, harbor the most massive and enigmatic objects in the universe: supermassive black holes. These cosmic monsters are incredibly dense regions of space that can weigh billions of times more than the sun, exerting an overwhelming gravitational pull that affects nearby stars' orbits and the distribution of matter within the galaxy. However, their influence goes beyond just shaping the galaxy's structure. They can cause catastrophic events that ripple through space. y absorbing matter in their vicinity, they can create accretion disks that discharge colossal quantities of energy through radiation and particle streams, transforming the destiny of the galaxies they inhabit. The scientific community is actively investigating these massive entities, hoping to unravel the enigmas of their origin, expansion, and conduct.

224. Existence of Supermassive Black Hole Binaries. Supermassive black hole binaries are systems comprised of two supermassive black holes orbiting each other. They are thought to reside at the centers of most large galaxies and can have masses up to billions of times that of the sun. The presence of supermassive black hole binaries offers insights into the formation and evolution of galaxies and can help study strong gravitational interactions and test general relativity in extreme environments. What is considered normal for one entity may not be so for another.

225. Elusive Intermediate-Mass Black Hole: A Key to Unlocking the Secrets of the Universe. Intermediate-mass black holes are a rare type of black hole with masses ranging from 100 to 100,000 solar masses, heavier than stellar-mass black holes but lighter than supermassive black holes found at the centers of galaxies. Confirmation of their existence has been elusive, with several false or ambiguous detections, but recent discoveries have provided strong evidence. Understanding intermediate black holes is of particular interest as they could provide clues on how black holes evolved over time and how supermassive black holes formed in the early universe. Their rarity is connected to the mystery surrounding the evolution of these objects, and the discovery of more intermediate-mass black holes could provide valuable insights into the formation of supermassive black holes.

226. Intermediate-Mass Black Holes in Dwarf Galaxies and Star Clusters. Recently, an international team of scientists found an intermediate-mass black hole in a dwarf galaxy during the Young Supernova Experiment scientific program, which they believe tore apart a star, an unusual object of intermediate mass. Meanwhile, astronomers at the University of Utah have confirmed the existence of an intermediate-mass black hole hidden within B023-G078, a star cluster located in the Andromeda galaxy, which is larger than black holes caused by supernovae but not as large as those found at the centers of galaxies.

227. **R**evealing the Unseen: Black Holes Are Invisible to the Naked Eye. Despite being invisible themselves, black holes can be detected through their gravitational effects on nearby matter circling them, such as stars or gas clouds. For example, when a black hole is in a binary system with a visible star, the star's gravitational pull can cause it to wobble or move in an unusual way. Another method black holes can be seen indirectly is through a phenomenon known as stellar motion. Additionally, black holes can be discovered by accretion disks, which frequently form around them. As matter falls into the black hole, it heats up and emits radiation, which X-ray telescopes can spot. Finally, gravitational sensing is used. When two black holes collide, they create gravitational waves, which are ripples in spacetime. Highly sensitive detectors, such as LIGO, can spot these waves. However, a black hole is completely undetectable.

228. **T**he Accretion Disk: Matter Around a Black Hole. An accretion disk is a flattened, circular, or elliptical structure that forms around a black hole as matter falls into it, creating intense radiation that telescopes can observe. This disk is formed when material falls towards a strong gravitational force, such as a black hole or a star. Accretion disks can be found surrounding various celestial bodies, from relatively small regions of a few thousand kilometers around white dwarfs and neutron stars to protoplanetary disks around very young stars. By analyzing the properties of the radiation emitted from an accretion disk, scientists can study the properties of the black hole at the center.

229. **W**hat **Would Happen If You Fell into a Black Hole: Strange Optical Distortions, Tidal Forces, and Doom.** If you fell into a black hole, once you pass the event horizon, you would be subjected to tremendous tidal forces that would stretch and squash you in different directions until you looked like a piece of spaghetti. The singularity, a point of infinite density at the center of the black hole, is still a mystery in terms of physics. However, before you reached the singularity, as you approached the event horizon, you would experience strange optical distortions of the surrounding sky due to the bending of light. For ordinary black holes, the tidal forces are so large that you would be stretched and squashed well outside the event horizon, likely killing you before you even reach the black hole. But for larger black holes, over 1000 solar masses, you might survive the fall into the black hole without even noticing the tidal forces until you pass the event horizon and become doomed.

230. **G**ravitational **Waves Confirm Black Holes Cannot Decrease in Size, According to Hawking's Area Law.** Physicists have confirmed Stephen Hawking's black hole area theorem, which states that a black hole's surface area cannot decrease over time. This law is related to the second law of thermodynamics, which states that the entropy of a closed system must always increase. The researchers used gravitational waves from two merging black holes to calculate their mass, spin, and surface area before and after the collision, confirming Hawking's area law with a 95% level of confidence. This discovery has challenged traditional theories and highlighted the need for a unified theory that can reconcile the principles of quantum mechanics and general relativity.

231. **T**he **Cosmic Symphony of Gravitational Waves.** Gravity waves are ripples in spacetime that propagate through the universe, created by massive objects such as black holes or neutron stars. These waves can cause distortions in the fabric of spacetime, and scientists can detect them by observing their effects on objects in space. Studying gravitational waves can provide valuable insights into the nature of black holes and other exotic objects and the fundamental nature of gravity itself.

232. **T**he **Ripple Effect: Einstein's Prediction Confirmed, The Revolutionary Discovery of Gravitational Waves.** The discovery of gravitational waves, predicted by Albert Einstein 100 years ago and confirmed by scientists earlier this year using the Laser Interferometer Gravitational-Wave Observatory (LIGO), marks the beginning of a new field of study known as gravitational wave astronomy. The detection of these waves was made possible thanks to Rainer Weiss's interferometer scheme, which proposed detecting them through optical instruments. Although the success of the LIGO team in detecting gravitational waves fulfilled the prediction made by Einstein, the discovery marks the beginning of a new era of exploration into the cosmos, and physicists are eagerly anticipating further discoveries in this new field. The next step is to spot many more events, including black hole mergers, which could be detected once a day if LIGO reaches

its design sensitivity. The upgraded VIRGO detector in Italy is also joining the hunt, and there is much excitement about what gravitational wave astronomy may reveal in the future.

233. **The Paradox of Black Holes: Challenges of Integrating General Relativity and Quantum Mechanics.** Black holes are one of the most enigmatic objects in the universe, posing unique challenges to our understanding of physics. While general relativity theory accurately describes black holes, the integration of general relativity and quantum physics remains a mystery to physicists. Quantum gravity, a field of science that seeks to reconcile quantum mechanics and general relativity, encounters difficulties when dealing with black holes. Black holes violate Hawking's area law, which says that the surface area of a black hole cannot decrease over time. However, Hawking's theory of Hawking radiation suggests that black holes can shrink and eventually evaporate over a long period of time due to quantum mechanics, violating the area law over long periods of time. The information paradox is one of the primary concerns of quantum gravity, questioning the fate of information that enters a black hole and leaving physicists grappling to understand the quantum behavior of these elusive objects. Despite the challenges, researchers will continue to analyze data from gravitational waves to gain further insights into black holes and hopefully shed light on the mysteries of the universe.

234. **The Biggest Black Hole Ever Found: TON 618.** TON 618, a supermassive black hole, is the largest black hole ever found, with a mass of 66 billion times that of the Sun and a size of 32.5 times wider than our Solar System. It powers a quasar that emits light estimated to be 10.37 billion years old, making it both large and ancient. Some scientists believe a black hole larger than TON 618 may not exist.

235. **T**he Fastest Growing Black Hole Ever Found Devouring an Earth-Size Object Every Second: SMSS J1144. A black hole, with a mass three billion times that of the sun, consumes the mass equivalent of an entire Earth every second, causing it to grow rapidly. This process is known as accretion, where matter is siphoned from a disk of gas and dust rotating around the massive object. The black hole is now 500 times larger than Sagittarius A*, the supermassive black hole at the heart of the Milky Way, and it is still growing. It is the most rapidly growing black hole found in the past 9 billion years and is powering an ultra-bright quasar. This quasar, designated SMSS J114447.77-430859.3, is the most luminous ever observed and has a brightness of magnitude 14.5 when viewed from Earth.

236. **T**he Fiery Spectacle of Relativistic Jets: A Black Hole Firing Fiery Jet at a Neighboring Galaxy. A team of astronomers has discovered a black hole emitting a fiery jet in another galaxy. The black hole is located in a galaxy called RAD12, which is situated around one billion light years away from Earth. Galaxies can be classified based on their appearance, with the two main categories being spirals and ellipticals. Spirals have spiral arms, while ellipticals appear yellowish and lack spiral arms. Spirals are more favorable to star formation, whereas ellipticals are uncommon. This is the first time a jet has been recorded colliding with a big galaxy like RAD12-B, and it could help scientists better understand the impact of such interactions on elliptical galaxies, which have limited star formation due to supermassive black holes depleting cold gas and dust.

237. **T**he Potential Role of Supermassive Black Holes in Dark Energy and the Universe's Expansion. The mysterious relationship between black holes and dark energy still fascinates scientists, as a new study suggests that supermassive black holes at the centers of galaxies may be the mysterious cause of the accelerating expansion of the universe. However, the notion is still being debated among specialists since the nature of black holes and their development in the early cosmos remain mysteries. Despite this uncertainty, it is possible that dark energy is confined within the interiors of black holes, and the mass of supermassive black holes should vary as the universe expands if this theory holds true. Remarkably, observations of elliptical galaxies have revealed that the black holes at their centers grew in mass without affecting their surrounding environments, offering further support for the notion that black holes could be responsible for dark energy. Nonetheless, skeptics remain unconvinced and insist on the need for additional evidence before accepting this paradigm-shifting concept. Regardless of the outcome, the possibility of revising fundamental assumptions about the workings of the universe underscores the importance of continued exploration and inquiry.

238. **Q**uasi-periodic Oscillations: Black Holes Emit X-rays in a Regular Pattern. Black holes emit X-rays, including high-energy "hard" X-rays, as gas spirals toward them through an accretion disk, heating up to temperatures of up to 18 million degrees Fahrenheit (10 million degrees Celsius). The motion of matter in the accretion disk around a black hole produces quasi-periodic oscillations (QPOs) that emit X-rays in a regular pattern, providing insights into the properties of the black hole, such as its mass and spin, and revealing the behavior of matter in extreme conditions near a black hole. The rhythmic palpitations of QPOs bear similarities to the human heart.

239. **Sagittarius A*: The Central Black Hole of the Milky Way.** Sagittarius A*, the central black hole of the Milky Way, weighs about 4 million times more than the sun and is situated 25,000 light-years away from Earth. As it slowly consumes nearby stars and gas, it grows even larger over time. Although it is an astonishing cosmic phenomenon, it is wiser to recognize the inherent value of belonging to the human race rather than those of neighboring stars.

240. **The Black Hole Unveiled: First-Ever Images of a Black Hole.** The first-ever images of a black hole, captured by the Event Horizon Telescope in 2019, revealed a bright ring-like structure surrounding the event horizon—the point of no return, where the gravitational pull is so strong that nothing, not even light, can escape. This unprecedented achievement not only confirmed the existence of black holes but also provided new insights into their behavior, such as the jet streams of superheated particles that spew out from their poles at nearly the speed of light.

241. **The Supermassive Black Hole at the Heart of the M87 Galaxy.** The M87 galaxy, located about 55 million light-years from Earth, is home to one of the universe's most massive and enigmatic objects: a supermassive black hole. This monster, located at the galaxy's center, has a mass of around 6.5 billion times that of the Sun and is surrounded by a massive accretion disk of gas, dust, and other debris. The intense gravity of the black hole is so strong that it warps the fabric of space-time

around it, creating a mesmerizing effect known as a "black hole shadow."

242. **The Existence of Binary Black Holes: A Dance Towards Cosmic Unison.** In 2016, the LIGO observatory confirmed the existence of Binary Black Holes (BBH) by detecting gravitational waves emitted by two merging black holes in a system known as GW150914. Binary stars undergo various physical processes during their lifespan, such as mass transfer, common envelope, tides, and natal kicks. Mass transfer and common envelope are crucial for the evolution of massive stars. After the main sequence, the star can expand, causing Roche lobe overflow and common envelope. If a black hole and a giant companion star undergo common envelope, they orbit each other surrounded by the giant's envelope, leading to a loss of kinetic energy and transfer of thermal energy to the envelope. If the envelope is ejected, the binary system becomes a BBH with a short orbital period. However, if the envelope is not ejected, the binary system merges prematurely, resulting in a single black hole. Unfortunately, our understanding of common envelope is still limited, which affects our knowledge of BBH demographics.

243. **The Cosmic Stretch: The Phenomenon of Spaghettification in Black Holes.** Spaghettification, also referred to as the "noodle effect," occurs when an object encounters an intense gravitational field, typically found around a black hole. Due to the extreme tidal forces within a black hole's singularity, any object that crosses its event horizon, such as an astronaut, star, or even light, is unable to escape its pull. As a result, the object expands vertically and horizontally under strong tidal forces, resembling the shape of a spaghetti noodle, hence the name "spaghettification".

244. **Y**oung Galaxies Emitting Extreme Radiation: The Fascinating World of Quasars. Quasars, initially called quasi-stellar radio sources, appeared star-like to astronomers in the late '50s and early '60s. However, they are not stars but rather young galaxies that exist at immense distances from us. As we move toward the edge of the visible universe, their numbers increase. Despite their great distance, quasars remain visible due to their extreme brightness, which can be up to 1,000 times greater than that of our Milky Way galaxy. They emit immense radiation, which makes it evident that they are highly active. Because quasars are so far away, we observe them as they existed in the early universe. Currently, the oldest known quasar is J0313-1806, situated around 13.03 billion light-years away. This means that we observe it as it was only 670 million years after the Big Bang.

245. **T**he Brightest Objects in The Universe: Quasars. Quasars are incredibly luminous, active galactic nuclei among the cosmos' most distant and ancient objects. Quasars are believed to be powered by the accretion of matter into supermassive black holes. Some quasars emit more light than an entire galaxy, equivalent to over 10^{14} times the energy of the Sun. Paradoxically, the brightest things in the universe exist thanks to the darkest things in the universe—black holes!

246. **R**are Double Quasars Offer Unique Insight into Supermassive Black Hole Collisions in Early Universe. Recently, a pair of quasars were discovered in the cores of two merging galaxies, indicating their proximity to each other. This rare discovery, along with the finding of another quasar pair in another colliding galaxy duo, provides a unique opportunity to study the collisions and merging of supermassive black holes in the early universe. Research estimates that for every 1,000 quasars in the distant universe, there is one double quasar. The discovery of these double quasars provides an exceptional chance to investigate the merging of supermassive black holes in the early universe.

247. **T**he **Largest Reserve of Water in the Universe: A Quasar.** Astronomers have discovered a massive cloud of water vapor located about 30 billion miles (48.28 billion kilometers) away in a quasar called APM 08279+5255, located in the constellation of Draco. This quasar is known for having the largest known reserve of water in the universe, with a cloud of water vapor surrounding it equivalent to 140 trillion times the mass of all the water in the world's oceans, along with other chemicals and molecular hydrogen. This discovery offers new insights into the potential for water and the universe's life-building blocks, suggesting that water, a key ingredient for life as we know it, might be more prevalent in the universe than previously thought.

248. **G**ravitational **Lensing: Observing Distant Galaxies.** Gravitational lensing is a phenomenon that occurs when the curvature of spacetime around a massive celestial body causes the path of light around it to be visibly bent, creating a lens-like effect. This effect is a prediction of Einstein's general theory of relativity, which posits that gravity is the curvature of spacetime. Gravitational lenses are often caused by massive objects such as galaxy clusters, which significantly curve spacetime. When light from a distant object passes through a gravitational lens, the path of the light is bent, resulting in a distorted image of the object.

"

The Earth is the cradle of humanity, but one cannot remain in the cradle forever.

– Konstantin Tsiolkovsky

SECTION 18
The Search for Habitable Planets and Signs of Life

249. **The Goldilocks Zone: Where Temperature is Just Right.** The Goldilocks zone, named after the fairy tale character who likes things to be "just right," is the region around a star where the temperature is suitable for liquid water on a planet's surface. This zone is also known as the "habitable zone" and is crucial for the potential of life to evolve. For a planet to be able to support life, it must not be too cold that water only exists as frozen ice, and it cannot be too hot that the water all boils away. Only planets within this specific range of orbits are believed to be capable of supporting life. In simpler terms, if a planet is too close to its star, it will be too hot, and if it is too far away, it will be too cold. The distance between a planet and its star must be just right for water to remain liquid, and this distance is the Goldilocks zone.

250. **Expanding the Search: Scientists Broaden Parameters for Extraterrestrial Life.** The search for extraterrestrial life has undergone a paradigm shift as scientists have broadened the parameters of their search to include the study of extremophiles, microorganisms that possess extraordinary resilience to extreme environmental conditions. This has led to reevaluating the potential for life to exist in places previously considered inhospitable, expanding the scope of the search beyond the traditional confines of our understanding.

251. **The Building Blocks of Life: The Search for Habitable Planets.** Scientists seeking life beyond Earth focus on three key indicators of a habitable planet: water, organic molecules, and energy sources. These indicators are essential because they constitute the building blocks of life, composed of six major elements: sulfur, phosphorus, oxygen, nitrogen, carbon, and hydrogen. The presence of these elements is a significant factor in the search for life on other planets. Even though these requirements may seem basic, their discovery remains a daunting challenge for astronomers and astrobiologists.

252. **P**otential **Candidates for Harboring Life.** The potential candidates for harboring life within our solar system include Mars, Europa, Enceladus, and Titan. These worlds have been found to have the necessary ingredients for life, such as water, energy, and organic molecules.

253. **E**uropa: **The Icy Moon and Its Hidden Ocean.** Europa, one of Jupiter's moons discovered in 1610 by Galileo, has a subsurface ocean estimated to be around 100 km deep. The Voyager 1 and 2 spacecraft provided the first detailed images of Europa's surface in 1979, revealing dark bands that suggest past activity and a lack of large impact craters, indicating a young surface. Scientists believe there may be a layer of liquid or warmer ice present between the outer crust and the deeper interior, as confirmed by the Galileo mission. Europa's ocean is thought to have more than twice the amount of water than all of Earth's oceans combined, and its liquid state is maintained by Jupiter's tidal flexing. The Europa Clipper and proposed Europa Lander missions aim to study the ocean's habitability, making Europa an intriguing candidate for the existence of extraterrestrial life.

254. **T**he Most Reflective Object in the Solar System: Enceladus. Enceladus, the sixth-largest natural satellite of Saturn, holds the distinction of being the most reflective object in the Solar System, with an albedo of nearly 100%. Its high reflectivity results in a surface temperature of approximately -329.8 degrees Fahrenheit (-201 degrees Celsius). Despite the frigid temperatures, Enceladus is not entirely inactive, exhibiting geysers and other geologic activity. The surface of Enceladus is blanketed by a layer of pure, pristine ice, contributing significantly to its high reflectivity. The exceptional brightness of this icy moon is of great interest to scientists and astronomers, shedding light on the conditions necessary for the existence of life and providing insights into the evolution and formation of our solar system.

255. **E**nceladus: A Small Moon with a Global Ocean of Liquid Water and Promising Signs of Life. This active moon of Saturn spews plumes of water into space from a global, liquid ocean hidden beneath its fissured crust. Cassini's mass spectrometer found a variety of organic molecules, building blocks of life, in the plumes, along with molecular hydrogen, carbon dioxide, methane, and rock fragments. Recent research shows that the plumes contain higher amounts of hydrogen, carbon dioxide, and methane than previously estimated, highlighting the importance of studying them to learn more about the subsurface ocean's composition and conditions that could support life. Enceladus also constantly spews frozen particles mixed with water and essential organic compounds, leading scientists to believe it may have an internal energy source similar to Europa. The potential for life on this small, active moon makes it an enticing target for exploration.

256. **Enceladus: A Moon with Active Cryovolcanism Erupting Volatile Substances into Space.** Enceladus is one of the most geologically active objects in the solar system, with evidence of ongoing cryovolcanism, or the eruption of volatile substances such as water and methane. The cryovolcanism on Enceladus is thought to be driven by heat generated by the decay of radioactive isotopes in its core, which causes the subsurface ocean to be in a state of constant motion. Scientists are fascinated by the cryovolcanism on Enceladus and are eager to learn more about its implications for the evolution and habitability of this intriguing moon.

257. **Titan's Atmosphere: A Thick and Unique Mixture of Nitrogen and Hydrocarbons.** Titan, Saturn's largest moon, is the only moon in our solar system with a significant atmosphere, which is primarily composed of nitrogen. However, unlike Earth's atmosphere, Titan's atmosphere also contains hydrocarbons such as methane and ethane. Despite its surface pressure is 50 percent higher than Earth's, Titan shares some similarities with our planet, such as having clouds, rain, rivers, lakes, and seas. Titan's atmosphere is a unique and mysterious feature of this remarkable celestial body.

258. **Titan's Liquid Hydrocarbon Lakes: A Unique Window into the Moon's Geology.** Titan, Saturn's largest moon, is home to stable liquid hydrocarbon lakes, such as methane and ethane, which make it unique in our solar system besides Earth. These lakes provide valuable information about the moon's geology, climate, potential for prebiotic chemistry and habitability, and conditions that may have existed on early Earth. The study of these lakes also opens the possibility of future exploration and

resource utilization, with potential applications for human space travel and settlement. The liquid hydrocarbon lakes on Titan are a fascinating and unique feature of this celestial body.

259. **D**iscovering Exoplanets. The Transit Method. The transit method is a strategy for finding exoplanets, or planets outside our solar system. The technique entails searching for small dips in a star's brightness as a planet passes in front of it, blocking some of the star's light. Astronomers can determine the planet's size and orbit by measuring the size and timing of these dips. Using the transit technique, thousands of exoplanets have been discovered, revolutionizing our understanding of the universe and the prevalence of other potentially habitable worlds.

260. **T**he Diversity of Exoplanets: The Possibilities of Extraterrestrial Life. The discovery of 51 Pegasi b in 1995 by Mayor and Queloz was unexpected, as theoretical models at the time suggested that gas giants could not form at such a short distance from the host star. This discovery opened up a new field of research in the search for extrasolar planets, leading to the discovery of thousands of exoplanets and planet candidates awaiting confirmation. The diversity of exoplanets is the most remarkable discovery in this field, challenging theoretical models and providing valuable insights into the possibilities of extraterrestrial life. Exoplanets offer scientists the chance to think about many scenarios for what alien life may look like. These worlds range from hazy atmospheres to frozen surfaces and anoxic conditions. One such exoplanet is Titan, Saturn's largest moon, which has the potential for life, a thick atmosphere, methane oceans, and other characteristics that set it apart from other celestial bodies. Future exploration and resource exploitation are also made possible by our understanding of Titan's liquid hydrocarbon lakes.

261. **More than 5,000 Confirmed Exoplanets.** As of 1 April 2023, there are 5,346 confirmed exoplanets in 3,943 planetary systems. Out of these systems, a significant number, 855, have multiple planets. Most of these exoplanets were discovered through the efforts of the Kepler space telescope, which has been instrumental in increasing our understanding of planetary systems outside of our solar system. It is pertinent to acknowledge the invaluable contribution of Kepler in expanding our perspective of the universe by revealing that our solar system is but a minuscule fragment within its vast expanse.

262. **The Search for a New Earth: A High Count of Potentially Habitable Exoplanets in Our Galaxy.** Exoplanets, planets that orbit stars other than our sun, have been discovered using various methods, and some of them are like Earth in terms of size, orbital distance, and temperature, raising the possibility of supporting life. The study of exoplanets has the potential to answer questions about the existence of life in the universe. According to data from the Kepler space mission, there could be as many as 40 billion Earth-sized planets orbiting the habitable zones of sun-like stars and red dwarfs in our galaxy, the Milky Way, and 11 billion may be orbiting sun-like stars. This high count of potentially habitable exoplanets makes the search for a new Earth an active and exciting field of research and raises the question of whether their wireless connectivity surpasses our own, which seems trivial in comparison.

263.   **The Moon's Role in Enabling Life Beyond Earth: Moons Matters in the Search for Habitable Planets.** The presence of a moon or moons can be substantial for the development and sustainability of life on planets in our universe. The moon's gravitational pull can help regulate a planet's rotation and stabilize its axial tilt, leading to a more stable climate that can support life. Additionally, the presence of a moon can influence tides and other geological processes that can impact the evolution of life. Thus, the presence of a moon is a significant factor to consider when searching for habitable planets and potential life beyond Earth.

"

The universe is under no obligation to make sense to you.

– Neil deGrasse Tyson

SECTION 19
Exoplanets: A Journey Through Eccentricity

264. **The First Exoplanet Discovered: PSR B1257+12 b.** In 1992, a team of astronomers led by Aleksander Wolszczan made the first discovery of an exoplanet around a pulsar star, PSR B1257+12, located 2,300 light-years away in the constellation Virgo. The discovery was made using the timing method, which involves observing variations in a pulsar's pulse arrival time due to the gravitational influence of an orbiting planet. The team found two exoplanets, PSR B1257+12 b, and c, with masses like the Earth's moon, orbiting the pulsar in 67 and 98 days, respectively. The discovery of exoplanets around a pulsar star was significant because it showed that planets could form in unexpected environments and around different types of stars.

265. **Exomoons: Elusive Natural Satellites Orbiting Exoplanets.** An exomoon is a natural satellite, or moon, that orbits an exoplanet. Exomoons have not yet been directly observed, but their existence is inferred based on observations of the exoplanets they orbit. Detecting exomoons is challenging because they are typically much smaller than their host planets, making it difficult to observe them directly. However, scientists have proposed several methods for detecting exomoons, such as measuring the slight variations in the timing and duration of the planet's transits or observing the moon's gravitational influence on the planet's orbit. The discovery of an exomoon would be significant because it could provide insight into the formation and evolution of planetary systems and the conditions necessary for life to arise. We can finally declare we're not alone in the lunar department.

266. **K**epler-444: An Ancient Star System with Rocky Planets Challenges Habitable Zone Theory. Kepler-444 is a unique star system located about 117 light-years away from Earth. It is composed of five small, rocky planets that are similar in size to Earth, and orbit the star at a much closer distance than Mercury orbits our sun. The most interesting fact about Kepler-444 is its age. The star is estimated to be about 11.2 billion years old, which means it is more than twice as old as our sun and one of the oldest known star systems in the galaxy. The discovery of the five tightly packed terrestrial planets orbiting such an ancient star has important implications for the possibility of life beyond our solar system. The fact that these planets formed so early in the universe's history suggests that there may be many more Earth-sized planets and that they may be more common than previously thought. This planet system is also interesting because it challenges the idea that rocky planets can only form in the habitable zone of a star, where conditions are just right for liquid water to exist on the surface. The planets in this system are too close to their star to be in the habitable zone, yet they are still rocky and Earth-like in composition.

267. **P**otentially the First Confirmed Exomoon: The Excitement Around a Neptune-Sized Moon. The potential discovery of an exomoon orbiting Kepler-1625b, a distant exoplanet in the constellation Cygnus, has caused excitement in the scientific community. This exomoon, which is thought to be roughly the size of Neptune, was discovered by NASA's Kepler satellite telescope when researchers saw a slight dilution in the star's brightness just before and after the planet's transit. Although the exomoon's existence has not yet been conclusively proven, it has opened up fresh research directions and improved

our comprehension of the intriguing dynamics of planetary systems.

268. **T**he Discovery of the First Exoplanet Orbiting a Sun-Like Star: 51 Pegasi b. The discovery of 51 Pegasi b in 1995 was a groundbreaking achievement in the search for exoplanets and marked the beginning of a new era in astronomy. The radial velocity method used by Michel Mayor and Didier Queloz to find this first exoplanet involved detecting the wobble of a star caused by the gravitational pull of an orbiting planet and has since become a widely used technique in exoplanet detection. The discovery of 51 Pegasi b challenged the traditional understanding of planetary formation and sparked new questions about the diversity of planetary systems in our galaxy.

269. **K**epler-186f: A Potential Haven for Life Beyond Our Solar System in the Habitable Zone. Kepler-186f is the first Earth-sized exoplanet discovered within its star's habitable zone. It orbits a red dwarf star approximately 500 light-years away in the constellation Cygnus and is believed to have a solid, rocky surface. The planet's position at the outer edge of the habitable zone means that it may have conditions suitable for liquid water, which is crucial for life. Kepler-186f's year is about 130 Earth days long and may be tidally locked. Its star is smaller and cooler than our sun, which could make the planet more hospitable to life, but also more vulnerable to space weather events.

270. **T**he Diamond Planet: 55 Cancri e. 55 Cancri e, the "super-Earth" exoplanet orbiting 55 Cancri A, has a composition primarily made up of carbon, potentially forming diamonds.

It's one of the closest exoplanets ever discovered, orbiting its host star every 18 hours. This planet's stark contrast between its scorching hot and freezing cold sides due to tidal locking, with temperatures on the hot side estimated to be over 3,600 degrees Fahrenheit (2,000 degrees Celsius), makes it an inhospitable environment for life as we know it. However, the unique conditions on 55 Cancri e have sparked fascination and curiosity about the possibility of alternative forms of life thriving in such extremes.

271. **The Most Complex Ring Systems Ever Found: The Real Lord of The Rings: J1407b.** This peculiar planet is located about 400 light-years away in the constellation Centaurus. It is known as the "ringed planet" due to its enormous ring system, which is thought to be up to 200 times larger than Saturn's rings, with 30-odd rings. What is not like Saturn, though, is the genuinely colossal size of the rings, which span 112 million miles (180 million kilometers) wide. That's larger than the Earth-Sun distance of 93.2 million miles (150 million kilometers) and 200 times bigger than Saturn's rings. The ring system is so massive that it blocks out a significant amount of light from its host star, making it possible to detect the rings using the transit method. It is considered one of the most complex ring systems ever found.

272. **The Blackest Exoplanet: HD 149026b.** HD 149026b is a gas giant exoplanet located approximately 256 light-years away in the constellation Hercules. It is notable for being the first exoplanet ever discovered to have an atmosphere, and its atmosphere was later found to contain water vapor. HD 149026b is a hot Jupiter with a mass about 0.36 times that of Jupiter and a radius about 1.3 times that of Jupiter. It orbits

its host star every 2.9 days at approximately 3.3 million miles (5.310.714 kilometers). The temperatures on HD 149026b range from 2,000 degrees Fahrenheit (1,093 degrees Celsius) to 2,800 degrees Fahrenheit (1,537 degrees Celsius).

273. **T**he Hottest Planet Ever Discovered: The Inferno KELT-9b. KELT-9b is a gas giant exoplanet that orbits very close to its host star, KELT-9, one of the hottest stars known. It is located approximately 670 light-years away in the constellation Cygnus. KELT-9b's extreme temperature is one of its most interesting features, reaching a scorching 7,800 degrees Fahrenheit (4,300 degrees Celsius), making it one of the hottest exoplanets ever discovered. Due to its close proximity to its star, the planet is tidally locked, causing a very elongated shape with a bulge on the hot side. KELT-9b's atmosphere contains high levels of metals such as iron, magnesium, and titanium, as well as molecular hydrogen and helium, indicating significant atmospheric loss due to the intense radiation and stellar wind from its host star. The planet's extreme temperature and shape make it an interesting subject for studying atmospheric dynamics and planetary formation, particularly the effect of radiation and stellar wind on atmospheric loss.

274. **T**he Coldest Exoplanet Ever Discovered: WISE 0855-0714. WISE 0855-0714 is an intriguing exoplanet that is believed to hold the record for the coldest known planet beyond our solar system. It is located about 7.2 light-years away from us, and its temperature is estimated to be around -375 degrees Fahrenheit (-226 degrees Celsius). This makes it a particularly challenging world to study, as its frigid temperatures are thought to be below the threshold at which methane gas can condense and form clouds, which makes it nearly invisible to telescopes. Nevertheless, astronomers continue to study this planet in the hopes of unlocking more clues about the nature and diversity of planetary systems throughout the universe.

275. **Kepler-16b: A 'Tatooine Planet' with an Evaporating Atmosphere Orbiting Two Stars.** Around 200 light-years from Earth, in the constellation Cygnus, is the intriguing exoplanet known as Kepler-16b. In a similar manner to the mythical planet from the Star Wars universe, this gas giant, also referred to as a "Tatooine planet," orbits two stars, a K-type, and an M-type star. The two stars orbit each other every 41 days, while Kepler-16b orbits them. One of the most unique features of Kepler-16b is its extreme proximity to its stars, which is causing the planet's atmosphere to evaporate rapidly. This phenomenon, known as "atmospheric evaporation," has been observed on various exoplanets, including Kepler-16b. Despite losing its atmosphere at an alarming rate, Kepler-16b still has a significant amount of gas surrounding it, making it a remarkable object for astronomers to study.

276. **Vanishing Worlds: The Phenomenon of Atmospheric Evaporation in Exoplanets.** Atmospheric evaporation is a phenomenon where the upper atmosphere of an exoplanet is heated to extremely high temperatures, causing the gas molecules to escape into space. This process can occur when an exoplanet is located very close to its host star, allowing the intense heat and radiation to strip away the outermost layers of the atmosphere.

277. **Triple Sunsets in Alien Sky: The Fascinating Exoplanet HD 188753 Ab**. HD 188753 Ab is an exoplanet located approximately 149 light-years away from Earth in the constellation Cygnus.

One of the most intriguing features of this exoplanet is that it orbits a triple-star system. The planet itself is a gas giant, with a mass about 1.14 times that of Jupiter, and it completes an orbit around its host star every 3.34 days. However, the most remarkable aspect of this exoplanet's orbit is that it is tilted by an angle of about 30 degrees concerning the plane of the triple-star system. This is a highly unusual configuration for an exoplanet, and it has challenged astronomers' understanding of planetary formation and evolution.

278. **W**ASP-12b: A Dark World with a Voracious Appetite for Light. WASP-12b is a fascinating planet that has attracted the attention of astronomers due to its unique quality of absorbing light. The planet's atmosphere contains a high concentration of carbon, which makes it a very efficient absorber of light. Furthermore, its close proximity to its parent star causes its atmosphere to break apart into individual atoms, contributing to its low reflectivity. These factors combine to make WASP-12b a planet that appears very dark, almost as if it's "eating up" light.

279. **M**etal Rain and Temperature Extremes: The Bizarre World of WASP-76b. WASP-76b is an exoplanet in a world of extremes, orbiting its star in just 1.8 days. Its close proximity to the star causes extreme heating, with temperatures on the dayside reaching over 4,300 degrees Fahrenheit (2,365.56 degrees Celsius). The heat causes heavy metals like iron to evaporate, forming clouds that move to the cooler night side, where they fall as "metal rain." Due to the temperature difference, the rain vaporizes before reaching the surface. The planet is tidally locked, with one side always facing the star and the other in permanent darkness, resulting in a stark temperature contrast.

280. **E**xoplanet Defying Physics with Solid Ice at Roasting Temperatures: Gliese 436b. Gliese 436b, an exoplanet located in the constellation of Leo, situated approximately 30 light years away from Earth, is a celestial body that challenges the fundamental laws of physics. Despite orbiting its host star at a distance 15 times closer than Mercury's orbit around the Sun, the surface temperature of Gliese 436b reaches a scorching hot 822 degrees Fahrenheit (439 degrees Celsius), enough to melt most substances. However, the icy surface of this planet remains solid due to the immense gravitational force exerted on it by the planet's core. This force compresses the scant amount of water vapor in the planet's atmosphere into solid ice, preventing it from melting, despite the intense heat. This remarkable phenomenon allows Gliese 436b to retain its icy surface despite the blazing temperatures. Additionally, it is noteworthy that the planet completes one orbit around its star in two days and 15.5 hours.

281. **T**he Dark Cherry Blossom: The Pink Planet GJ 504 b. One of the most exciting facts about GJ 504 b is its unique color. The exoplanet appears to have a deep magenta hue caused by its young age and relatively low temperature. Scientists believe the planet is around 160 million years old, relatively young compared to most exoplanets. Another intriguing aspect of GJ 504 b is its size. The exoplanet is about four times the mass of Jupiter and is considered a "super Jupiter." Despite its large size, GJ 504 b orbits its star at a distance more than four times greater than the distance between Neptune and the Sun. This makes it one of the largest exoplanets to be discovered at such a large distance from its host star.

282. **The Waterworld of GJ 1214b: A Unique Exoplanet Dominated by Water.** GJ 1214b is known as the "Waterworld" exoplanet because of its unique composition. Scientists estimate that up to 75% of the planet is made up of water or other volatile compounds, such as methane or ammonia. Discovered in 2009 by the MEarth Project, GJ 1214b is located in the constellation Ophiuchus, about 40 light-years away from Earth. However, due to its proximity to its host star, GJ 1214b is subjected to high surface temperatures. The planet orbits a small, cool, red dwarf star with an orbital period of only 1.6 days, resulting in temperatures estimated to range from around 750 to 1,100 degrees Fahrenheit (400 to 600 degrees Celsius). The planet's dense atmosphere, combined with the high temperatures, creates a challenging environment for life as we know it to survive. Scientists believe that the water on the planet is likely in a supercritical state, neither liquid nor gas, but exists in a hot, dense, and highly pressurized state that is difficult to replicate on Earth. Further research is required to determine the exact properties of the water and other volatiles on this intriguing "waterworld" exoplanet.

283. **A Planet Where It Rains Glass Horizontally: HD 189733b.** HD 189733b is an exoplanet with clouds made of glass beads created by silicate rain. The planet is scorching, with a surface temperature of over 1,832 degrees Fahrenheit (1,000 degrees Celsius), and has a thick atmosphere composed mainly of hydrogen and helium. Strong winds up to 5,400 miles per hour (8,700 kilometers per hour) and high temperatures cause silicate particles in the atmosphere to melt and form droplets of liquid glass, creating a rain of glass beads.

284. **T**he Biggest Planet Ever Found: HD 100546b. This colossal exoplanet is located approximately 348 light-years away from the Earth in the constellation Musca and is considered one of the most mysterious and largest exoplanets ever discovered. HD 100546b is thought to be a young planet still in the process of forming, and its strange orbit and unusual behavior have puzzled astronomers for years. This massive world weighs about 762 Jupiter masses and orbits its star further than Earth orbits the Sun, and its radius is about 6.9 times that of Jupiter. The planet is also surrounded by a complex disk of gas and dust, which is thought to be the remnants of the planet's formation process. Despite its strange behavior, HD 100546b is considered a unique system that provides insights into the formation and evolution of planetary systems. These mysterious exoplanetary worlds challenge our understanding and make us question what we know.

285. **T**he Blue Giant: HD 189733b. A hot "Jupiter" exoplanet located about 63 light-years away in the constellation Vulpecula. It is considered one of the best-studied exoplanets and has been found to have a unique color—a deep blue. The blue color is caused by silicate particles in its atmosphere, which scatter blue light more efficiently than other colors. This suggests that exoplanets can have a wide range of colors, not just those seen in our own solar system.

286. **D**ense Giant Planet: High-Eccentricity Migration and Metal Enrichment. NASA's TESS has discovered a large planet orbiting a star called TOI-4603. This planet is much bigger than Earth, with a radius of about 1.04 times that of Jupiter. It takes around 7.25 days to complete one orbit. The planet's mass is about 12.89 times that of Jupiter, making it very dense. It lies in a rare region between massive giant planets and low-mass brown dwarfs. Its slightly elongated orbit and proximity to the star suggest high-eccentricity tidal migration. The planet also contains a significant amount of heavy elements and is enriched in metals compared to its star. Studying such systems can enhance our understanding of how massive planets evolve.

287. **Kepler-438b: The Eerily Earth-Like Exoplanet in the Constellation Lyra.** Kepler-438b is a rocky exoplanet located in the constellation Lyra, and it is considered one of the most Earth-like exoplanets found to date. It has a mass similar to Earth's and is located in the habitable zone of its star, making it a good candidate for potential habitability. It orbits a red dwarf star, and its orbit takes it close to its star, giving it a year of 35 days. It's so strikingly similar that it's almost eerie.

288. **Kepler-442b: The Remarkably Similar Earth-Like Planet Orbiting a Cool K-Type Star.** Kepler-442b is another exoplanet considered one of the most similar to Earth. It has a mass slightly larger than Earth, is located in its star's habitable zone, and has an orbital period of 112 days. This exoplanet orbits a K-type star that is much cooler and smaller than our sun. This world is a close parallel to our beloved planet.

289. **Proxima b: The Potentially Habitable Exoplanet Neighbor.** Proxima B is an exoplanet in the habitable zone of Proxima Centauri, the closest star to our Solar System at just 4.2 light-years away. With a mass of 1.17 times that of Earth and an orbital period of 11.2 days, it has the potential to support liquid

water on its surface and has been proposed as a potential destination for future manned missions. This exoplanet is exciting to scientists because of its proximity to Earth.

290. **T**RAPPIST-1: A Planetary System with the Potential to Host Life. TRAPPIST-1 is a planetary system that has been extensively studied using various ground and space telescopes such as Kepler, Spitzer, Hubble, and soon the James Webb Space Telescope. In 2017, seven Earth-sized planets orbiting TRAPPIST-1 were discovered, all of which are within the habitable zone and may have the potential to support life. Further research revealed that some of these planets may have more water than Earth, in different forms. A recent study found that the TRAPPIST-1 planets likely contain similar materials as rocky planets, including iron, oxygen, magnesium, and silicon. However, their density is about 8% lower than Earth's, indicating a different ratio of elements.

291. **P**lanetary Neighbors: TRAPPIST-1 System Planets Visible from Each Other's Surfaces. TRAPPIST-1's closely orbiting exoplanets may not have moons due to their proximity to the star. However, standing on the surface of one of these planets would provide a clear view of the other planets in the system. At certain times, these planets would appear larger than our moon because TRAPPIST-1 is only slightly larger than Jupiter, and its planets orbit at a similar distance to Jupiter's moons.

292. **E**xoplanet HAT-P-7b: The Planet Where It Rains Precious Gems. HAT-P-7b is a gas giant exoplanet lying in the HAT-P-7 system, approximately 1,000 light years away from our planet. It is considered one of the most unique exoplanets due to its

extremely high surface temperature, which reaches over 2,500 degrees Celsius. Scientists believe that the intense heat and pressure on HAT-P-7b's surface cause the formation of exotic materials, such as sapphires and rubies, which rain down from its atmosphere. The exoplanet orbits its star in just 2.2 Earth days and is considered inhospitable to life as we know it. The discovery of HAT-P-7b was made in 2008 through the transit method.

293. **The Icy Planet 22,000 lights Years Away: OGLE-2005-BLG-390Lb.** OGLE-2005-BLG-390Lb is a potentially icy gas exoplanet located in the OGLE-2005-BLG-390 system, approximately 22,000 light years away from Earth. It is considered one of the most distant exoplanets ever discovered and is believed to have a surface covered in frozen water and gas. The exoplanet orbits its star at a distance of over eight astronomical units, making it one of the most remote exoplanets in our search for life beyond our solar system. Despite its distance, the discovery of OGLE-2005-BLG-390Lb has given scientists valuable insights into the conditions and composition of distant exoplanets.

294. **A Planet Almost as Old as the Universe: PSR B1620-26 b.** PSR B1620-26 b is a gas giant exoplanet located in the PSR B1620-26 system, approximately 12.4 billion years old. It is considered one of the oldest exoplanets in the universe and is believed to have formed just one billion years after the Big Bang. PSR B1620–26 b has a mass of 2.6 times that of Jupiter and orbits a binary star system, making it one of the few exoplanets that orbit two stars. The exoplanet is considered inhospitable to life as we know it due to its age and location in a binary star system.

295. **W**ASP-17b: The Unusual Exoplanetary Migrant. WASP-17b is a gas giant exoplanet located in the WASP-17 system in the constellation Scorpius, approximately 1400 light years from Earth. It is considered one of the most unusual exoplanets ever discovered, as it orbits its star in just a few hours, much faster than any other known exoplanet. WASP-17b is also considered a "hot Jupiter" due to its size and proximity to its star. The origin of WASP-17b and its unusual orbit remain mysteries to scientists, who have yet to determine how the exoplanet formed and how it ended up in its close orbit. Despite its inhospitable environment, the discovery of WASP-17b has provided valuable insights into the diversity of exoplanetary systems and the processes that shape them. A mystery worthy of inclusion in your cabinet of curiosities.

296. **H**AT-P-7b: The Sapphire and Ruby-Raining Planet. HAT-P-7b is a gas giant exoplanet located about 1000 light-years from Earth. It is notable for its extremely high temperatures, with temperatures on its dayside reaching up to 4,600 degrees Fahrenheit (2,538.89 degrees Celsius). This extreme heat causes the atmosphere to vaporize heavy elements such as titanium and vanadium, condensing into sapphire and ruby clouds.

297. **T**he Unlikely Birth of Pulsar Planets: A Rare and Fascinating Phenomenon. Pulsars, which are formed from the collapse of a massive star during a supernova, are some of the most extreme and powerful objects in the universe. During this process, gravity pulls the mass of the star inward until it implodes without the opposing force of nuclear fusion to balance it. Although the formation of pulsars is a rare and

fascinating phenomenon, even rarer is the creation of pulsar planets. Despite the low probability of this occurrence due to the massive size of pulsars, there are several scenarios in which planets can form, such as when a planet survives a supernova, or when the destruction of a companion star creates a disk around the pulsar. Another possibility is the evaporation of a companion star into a planet-sized object due to the intense radiation emitted by the pulsar. However, the formation of pulsar planets remains a fascinating and largely unexplained phenomenon that continues to capture the imagination of astronomers and space enthusiasts alike.

298. **Pulsar Planets: Where Diamond Cores May Be the Norm.** Pulsar planets, which are a rare type of planet orbiting a pulsar, have been found to have unique features. They may have diamond cores due to the extreme pressure and temperature conditions in their interiors. The carbon atoms could be compressed into diamond crystals, leaving only the innermost part of the planet composed of crystalline carbon and oxygen. In 2011, scientists detected a pulsar planet with an estimated mass twice that of Jupiter, primarily composed of carbon and oxygen. The extreme conditions on these planets could cause the carbon to crystallize into diamonds, making these pulsar planets unique and possibly the only known planets in the universe with diamond cores.

299. **The Chthonian Planet: Born from the Ashes of its Parent Star.** Chthonian planets, also known as "stripped" planets, are a type of planet born from the ashes of their parent star. These planets are formed from gas giants that have migrated too close to their parent star, causing the star's heat and radiation to strip away their outer gas layers. What's left is a rocky core that resembles a super-Earth or Neptune-sized planet but with a mass and density much greater than what would be expected for a planet of that size. One of the most interesting things about Chthonian planets is that they offer a window into the early stages of planetary formation and evolution, providing insights into the physical and chemical processes that shape the universe. Furthermore, their proximity to their parent star makes them excellent targets for studying the atmospheres of exoplanets and searching for signs of life. Like the mythical phoenix, they are symbols of rebirth and transformation, proof that even in the face of adversity, life can find a way to endure and thrive.

300. **P**ulsar **Planets: Exoplanets that Orbit Around Pulsars.** Pulsar planets are a type of planet that orbits around pulsars, which are rapidly rotating neutron stars that emit beams of radiation that sweep across space like a cosmic lighthouse. These planets are unique in that they orbit stars that are remnants of supernova explosions rather than conventional stars. Pulsars are incredibly dense, with masses comparable to that of the Sun compressed into spheres just a few miles in diameter. Because of their extreme gravity, pulsars emit intense radiation that is deadly to life as we know it. It is just right for water to be present on the surface. The study of pulsar planets is still in its infancy, but it offers a glimpse into the strange and astonishing diversity of the universe.

301. **T**he **Rogue Planets: Interstellar Planets That Drift Alone in the Galaxy.** Rogue planets, also known as interstellar planets, are planets that do not orbit a star but instead drift through the galaxy alone. These planets can form through a variety of mechanisms, including gravitational interactions with other

objects and ejection from their original star systems. Rogue planets can have a wide range of temperatures and sizes, from small icy worlds to gas giants, and they may even have conditions suitable for supporting life. Experts believe there may be billions of rogue planets in the Milky Way galaxy alone, and they could outnumber stars. Rogue planets are difficult to detect, but the Nancy Grace Roman Space Telescope is expected to help scientists detect these wandering planets through microlensing. Studying rogue planets could provide insights into how planets form and evolve in the absence of a host star's influence, offering a unique opportunity to study planetary formation and evolution. While experts do not think there is a rogue planet in our solar system, learning about these lone wanderers in the galaxy could become very important to the future of our planet.

302. **The Lone Wanderer: The Discovery of a Rogue Planet with Auroras.** In 2019, scientists discovered a rogue planet, SIM J01365663+0933473, that is not gravitationally bound to any star. This planet is also unique because it has auroras similar to those seen on Earth and other planets in our solar system that are in close proximity to a star. The discovery of this rogue planet with auroras has essential implications for our understanding of planet formation and the potential for life on other planets.

303. **The Eccentric Planets: The Wild Ride Around Their Host Star.** Eccentric planets are celestial bodies that have orbits that are highly elongated and stretched out, causing their distance from their host star to vary greatly. This can result in extreme seasonal differences, with some areas of the planet experiencing long, hot summers while other regions face bitterly cold winters. Additionally, the gravitational forces acting on these planets tend to be stronger, often resulting in them being more massive than planets with circular orbits. Eccentric planets can also exhibit unique features such as backward-spinning moons and tilted rings, and studying them is essential to understanding planet formation and evolution. However, their extreme temperature fluctuations could make it difficult for life to exist, making the habitability of these planets an ongoing area of research.

304. **T**he **Binary Planets: A Double Delight in the Sky.** Binary planets are a type of planet system in which two planets orbit around a common center of mass. These planets can come in a variety of sizes and configurations, with some binary planet systems featuring two planets of roughly equal size and others featuring a large planet orbited by a smaller moon-like companion. One of the most interesting things about binary planets is the complex gravitational interactions that occur between the two bodies. These interactions can cause tides and atmospheric changes and even result in material exchange between the two planets.

305. **T**he **Enigmatic Planetary System in a Binary Star System: HD101065.** The HD101065 star system is a binary star system located about 330 light-years away from Earth. It is notable for its planetary system, which includes at least one planet (HD101065 b) that orbits one of the stars in the system. This planet is considered a "warm Jupiter," a gas giant like Jupiter but with an orbital distance closer to its host star, making it much hotter than Jupiter.

"

*As we look up to the stars,
we are confronted with the
infinitude of the universe,
a reminder that our human
concerns are but a speck in
the vast cosmic expanse.*

– Epicurus

SECTION 20
Stars: From Birth to Death

306. **T**he Life Cycle of Stars: From Formation to Supernovae and the Origins of Precious Metals. Stars are born when clouds of gas and dust collapse under gravity, forming dense cores that ignite nuclear fusion. They shine brightly for millions to billions of years, fusing elements in their cores until they exhaust their fuel. When a star runs out of fuel, it can either collapse into a neutron star or a black hole or explode in a supernova. More massive stars go through a phase of mass loss, either involving a stellar wind or a more violent explosion. Supernovae are responsible for the production of neutron stars and probably black holes, which recycle rare and important heavy elements into the universe. The cosmic life cycle, from star formation to supernovae and the origins of precious metals, is a continuous and interconnected process in the universe.

307. **U**Y Scuti: The Biggest Star of the Cosmos. The biggest star ever found in space is not just big; it's colossal. UY Scuti is a red supergiant that reigns supreme at a staggering 1,700 times the size of our own Sun. In comparison, our Sun is a mere speck in the grand scheme of things. But don't let its size fool you; UY Scuti may be large, but it's also near the end of its life and is expected to go supernova at some point. This is thought to occur within the next 100,000 years.

308. **8**8 Constellations: A Map to Navigate the Stars. Constellation maps divide the celestial sphere into 88 parts called constellations, aiding astronomers in locating stars and deep-sky objects. Visible star constellations depend on the observer's location and season, and they change throughout the year. The International Astronomical Union recognizes 88 constellations, with 48 of them having been recorded by the Greek astronomer Ptolemy in 150 AD. Stars appear to move from east to west as the Earth rotates on its axis, and as the Earth orbits the Sun, constellations shift westward throughout the year. Different parts of the sky become visible as the seasons change, enabling observers to view different constellations at different times of the year. There are three types of constellations: ancient, traditional, and contemporary. These constellations guide astronomers as they study the stars and investigate the night sky.

309. **S**tars Form Before Galaxies: A Cosmic Sequence. The birth of the first stars in the universe preceded the formation of galaxies. Hydrogen and helium gas clouds were the building blocks for the first stars, which eventually grouped to form the early stages of galaxies. This process took several billion years and resulted in the large and complex structures we observe today.

310. **Stars Twinkle, Planets Don't.** Stars twinkle because of atmospheric refraction, while planets don't. Stars appear to twinkle in the night sky due to the turbulent movement of the Earth's atmosphere, causing light from distant stars to be refracted and scattered in different directions before reaching our eyes. On the other hand, planets are much closer to Earth and appear larger in the sky, reducing the effect of atmospheric refraction and making their light appear more steady and less prone to twinkling. This phenomenon is known as scintillation and results from the varying air densities and temperatures that light passes through as it travels through the Earth's atmosphere.

311.	**T**he Brightest Star in the Night Sky: Sirius. Also known as the Dog Star, it is the most luminous star in the night sky. It is impressively visible during the winter night sky of the Northern Hemisphere and is relatively close to our home, Earth, just 8.6 light-years away. Sirius is a main sequence star, which means it is in the process of burning hydrogen to create helium. It is over 20 times brighter than our sun, for better understanding, and has an apparent magnitude of -1.46.

312.	**T**he Binary Secrets of Sirius: A Dynamic Star System. Sirius, a star located in the constellation Canis Major, is a binary star system composed of Sirius A and Sirius B. Sirius A is a white main-sequence star and the more massive of the two. At the same time, Sirius B is a white dwarf star. The two stars have a close orbital period of just 50.1 years and appear as a single point of light to the naked eye. Still, their proximity and difference in mass result in a strong gravitational interaction and matter transfer between the stars, making Sirius a unique and dynamic binary system. This discovery provided valuable insights into the evolution and behavior of binary star systems and remains one of the lesser-known but fascinating facts about Sirius.

313.	**W**hen Stars Go Boom: The Fascinating World of Blitzars. Blitzars are incredibly intriguing astronomical phenomena that offer insights into the fiery demise of massive stars. These rare objects are a blend of pulsars and black holes, resulting from the cataclysmic explosion that occurs when a massive star collapses into a black hole, flinging out its remaining matter in a burst of gamma-ray radiation visible across the universe. The detection of blitzars is complex, but scientists are captivated by their potential to unravel the secrets of black holes and dark matter. By studying these objects, we can better understand the most enigmatic phenomena of the universe, from the creation of black holes to the properties of dark matter.

314. **The Behavior of Stars in Extreme Environments Like Starburst Galaxies**. Starburst galaxies are extreme environments where star formation occurs at an extremely high rate, often hundreds to thousands of times faster than in our Milky Way galaxy. The stars in these environments are formed under different physical conditions than stars in less active galaxies. The intense radiation, supernovae, and massive stellar winds produced by the young, hot stars in starburst galaxies can affect the behavior of other stars in the galaxy and contribute to the overall evolution of the galaxy.

315. **Hypervelocity Stars: The Fastest Objects in the Milky Way.** Did you know there are stars in the Milky Way capable of ejecting material at incredibly high speeds? These stars, known as hypervelocity stars, are thought to be the result of a close encounter with a supermassive black hole in the center of our galaxy. During such an encounter, the gravitational pull of the black hole can accelerate a star to speeds greater than the escape velocity of the Milky Way. This causes the star to be flung out of the galaxy at speeds of approximately 1.1 million to 2.2 million miles per hour (1.8 million to 3.5 million kilometers per hour), making it the fastest object in the Milky Way. While hypervelocity stars are rare, their study provides important insights into the dynamics of the galactic center and the influence of black holes on their surrounding environments.

316. **S**hining Stars: **The Stellar Nuclear Fusion Reactions.** Stars are celestial bodies that generate light and heat through nuclear fusion reactions in their core, fusing hydrogen atoms to form helium and releasing massive amounts of energy in the process. The energy produced makes stars shine and provides warmth and illumination to the universe. Our Sun is just one of an estimated 100 billion stars in the Milky Way alone, illuminating the cosmos with its light and heat. However, stars also have a limited lifespan and eventually run out of fuel, leading to their collapse or explosion.

317. **S**hooting Stars: **The Spectacular Display of Meteor Showers.** Meteor showers are celestial events that bring a burst of activity to the night sky. They occur when the Earth passes through the debris trail left behind by a comet, resulting in a flurry of shooting stars illuminating the darkness. These awe-inspiring displays of light and motion can be seen from various locations on our planet and are a true testament to the beauty and power of the universe. It is like watching an epic Star Wars battle play out in the night sky.

318. **E**xploding Stars: **Supernovae Release Energy Equal to Several Suns.** Supernovae are a class of violently exploding stars that release tremendous amounts of energy, including radio waves, X-rays, and cosmic rays. They also release heavier elements into the interstellar medium, and their remnants can be observed as supernova remnants. Historically, only seven supernovae were recorded before the early 17th century, with the most famous being the one that occurred in 1054 and resulted in the Crab Nebula. The closest and most easily observed supernova was SN 1987A, which peaked in brightness

in May 1987. Supernovae can be divided into two broad classes, Type I and Type II, based on the way they detonate.

319. **T**he HR Diagram: Mapping the Evolution of Stars Based on Temperature and Luminosity. The Hertzsprung-Russell (HR) diagram is a fundamental tool in studying the evolution of stars. It illustrates the relationship between a star's temperature (or color) and luminosity (or absolute magnitude), revealing its internal structure and evolutionary phase. As stars undergo specific stages based on their mass, their positions on the HR diagram change accordingly. The main sequence, where stars are fusing hydrogen into helium, is the dominant region on the HR diagram. Above the main sequence, you'll find red giants and supergiants, which are burning heavier elements. On the other hand, white dwarfs, the final evolutionary stage for low to intermediate-mass stars, are located at the bottom left of the HR diagram. To differentiate between stars of the same temperature but different luminosity, Morgan-Keenan luminosity classes are utilized.

320. **B**lack Holes: The Ultimate Destiny of Some Stars. Black holes are formed when massive stars run out of fuel and collapse under their own gravity, becoming so dense that not even light can escape their gravitational pull. These celestial objects are the ultimate destiny of some stars, and their extreme gravity warps space and time around them, making them some of the most mysterious and intriguing objects in the universe.

321. **T**he Most Common Star in the Universe: Red Dwarfs. Red dwarfs are the most abundant type of star in the universe, making up approximately 75% of all existing stars, and having lifetimes of up to 10 trillion years. Despite their small size, they have enough mass to sustain nuclear fusion in their cores and shine for billions of years. The long lives of red dwarfs make them important contributors to the evolution of galaxies, as they continue to shine long after more massive stars have exhausted their fuel.

322. **F**rom Main Sequence to Red Giant: The Life Cycle of Stars. Red giant stars are a natural part of the life cycle of many stars, including our own Sun, which is predicted to become a red giant in about five billion years. Most stars in the universe are main sequence stars, which generate energy by converting hydrogen into helium in their cores through nuclear fusion. However, once this core fusion stops, gravity takes over and begins to compress the star. This causes the internal temperature of the star to rise, igniting a shell of hydrogen burning around the inert core, which in turn causes the star to expand and become a red giant.

323. **T**he Remnants of Stars: White Dwarfs. White dwarfs are the remnants of stars like the Sun that have exhausted all their fuel and collapsed to a size comparable to that of the Earth. They are incredibly dense and hot; indeed, they are one the densest objects in space, surpassed only by other compact stars such as neutron stars, black holes, and quark stars. Their temperatures can reach around 180,032 degrees Fahrenheit (100,000 degrees Celsius). As cosmic fossils, white dwarfs are the final remnants of stars that have lived and died, marking the end of a stellar journey that began billions of years ago. Their peculiar properties have made them important objects

of study, offering insights into the structure and evolution of stars.

324. **The Failed Stars: Brown Dwarfs - Bridging the Gap between Planets and Stars.** Brown dwarfs are mysterious objects that blur the boundary between planets and stars. They are too massive to be considered planets but do not have enough mass to sustain nuclear fusion in their cores, a defining feature of stars. These "failed stars" are cool and faint, and difficult to detect, making them a challenge to study.

325. **The Cosmic Ghost: The Possible Existence of Dark Stars.** Dark stars are a theoretical type of star-powered by dark matter instead of nuclear fusion. They are much cooler than common stars and do not emit visible light, making them invisible to telescopes. However, they would still generate heat, which could be detected by their infrared radiation. Scientists are still searching for their existence, but if discovered, they could help shed light on the nature of dark matter and its role in the universe.

326. **The Fiery Fate of Stars: Exploring the Red Giant Phase in the Life Cycle of Celestial Bodies.** Red giant stars are a natural part of the life cycle of many stars, including our own Sun, which is predicted to become a red giant in about five billion years. Most stars in the universe are main sequence stars, which generate energy by converting hydrogen into helium in their cores through nuclear fusion. However, once this core fusion stops, gravity takes over and begins to compress the star. This causes the internal temperature of the star to rise, igniting a shell of hydrogen burning around the inert core, which in turn causes the star to expand and become a red giant. Some red dwarfs may still be shining billions of years from now, long after the Sun and other stars have gone dark.

327. **The Colossal Giants: Hypergiants.** Among the rarest and most massive stars in the universe, hypergiants are behemoths that can have masses over 100 times that of the Sun and shine with millions of times their luminosity. Their short lifetimes of a few million years mean they are destined to end their lives in a spectacular supernova explosion. The immense size and power of hypergiants shape the surrounding interstellar medium, producing shockwaves that trigger the formation of new stars.

328. **The Extreme Heat: Blue Supergiants**. Among the hottest and most massive stars in the universe, blue supergiants shine with surface temperatures that can reach up to 90,032 degrees Fahrenheit (50,000 degrees Celsius). These stars consume their fuel at an astonishing rate, making them short-lived objects that live only a few million years before ending their lives in a spectacular supernova explosion. The immense heat generated by blue supergiants drives complex processes, such as stellar winds, that shape the surrounding interstellar medium and play a vital role in the formation of new stars.

329. **Supernova Types: The Explosive Deaths of Stars.** Supernovae are some of the most powerful and dramatic events in the universe, marking the explosive deaths of massive stars. There are several different types of supernovae, each with its own unique characteristics. Type Ia supernovae occur in binary star systems when a white dwarf star absorbs

matter from a companion star until it hits a critical mass, causing a runaway nuclear reaction that destroys the star. Type II supernovae are caused by the center of a massive star collapsing, creating a shockwave that blows off the star's outer layers. These occurrences are critical to comprehending the evolution of stars and the formation of heavy elements in the cosmos.

330. **The Cosmic Beacon: Superflares.** Superflares are sudden, massive outbursts of energy from stars that can release as much energy as the star emits over several days or weeks. They are prevalent in young, active stars, and their causes include magnetic reconnection and collisions of stellar material. These flares can significantly affect any exoplanet orbiting these stars, severely affecting their atmosphere and habitability.

331. **The Fascinating Life Cycle of Stars: Mass Determines Their Destiny.** Stars go through different stages of evolution, and their life cycle depends on their mass. The mass is determined by the amount of matter available in the nebula, the giant cloud of gas and dust from which the star is born. As the hydrogen gas in the nebula collapses under gravity, it forms a spinning protostar that eventually reaches a temperature of 15,000,000 degrees and undergoes nuclear fusion in the core, becoming a main sequence star. Our sun is currently in this stage, shining for millions to billions of years by converting hydrogen into helium through nuclear fusion in its core. However, as the hydrogen supply in the core runs out, the star becomes unstable, contracts, and expands, forming a red giant phase. All stars evolve in a similar way up to this phase, and their mass determines their subsequent life cycle path.

332. **P**owerful **Outflows from Death Star Galaxies.** Death Star galaxies are named for their capability to produce powerful galactic winds driven by the presence of a supermassive black hole at their center. These winds can significantly impact their surroundings and the evolution of other galaxies by removing large amounts of gas and dust from the galaxy, reducing star formation, and heating and ionizing the intergalactic medium. The study of Death Star galaxies provides valuable insights into the role of black holes in shaping the universe and the processes that influence galaxy formation and evolution. Even Darth Vader would be proud of these cosmic warriors.

333. **T**he **Mysterious and Dense World of Neutron Stars: Remnants of Exploded Supermassive Stars.** Neutron stars are extremely dense remains of supermassive stars that have undergone supernovae. The evolution and fate of a star depend largely on its mass, and all supermassive stars have the potential to become neutron stars. If the core of the star after the supernova explosion has a mass less than about three times that of the Sun, it forms a neutron star. Neutron stars are incredibly dense, with a mass greater than that of the Sun but a radius of only about 10 kilometers. They are mainly composed of neutrons, with most of the protons and electrons combined to form neutrons under dense conditions. Although they do not actively generate heat through nuclear fusion, neutron stars are incredibly hot.

334. **T**he **Strongest Material in Space: Nuclear Pasta.** Nuclear pasta is a hypothesized type of matter that may exist in the ultra-dense core of neutron stars. Neutron stars are formed when the core of a medium-sized star collapses and explodes in a supernova. The high pressure inside the core causes the neutrons to form unique structures called nuclear pasta. These structures are incredibly strong and take the form of various types of pasta such as gnocchi bubbles, spaghetti, and lasagna sheets. The density of nuclear pasta can be as high as a billion times greater than that of lead. The pressure inside a neutron star is so immense that protons and electrons are forced to combine to form neutrons, making nuclear pasta one of the strongest known materials in the universe. A sugar-cube-sized piece of nuclear pasta would weigh around a billion tons on Earth, and its density can reach up to 10^{17} kg/m^3, which is 10 billion times stronger than steel. Although nuclear pasta has not been directly observed, it is predicted to exist and is a topic of ongoing research in the field of astrophysics.

335. **M**agnetic Stars: Trillion of Times Stronger than Earth's Magnetic Field. Magnetic stars, also known as magnetars, are intriguing celestial objects with magnetic fields around 1,000 times stronger than a normal neutron star and a trillion times stronger than the Earth's. In fact, if you were to get too close to a magnetar, the magnetic field would be strong enough to tear electrons away from your atoms and convert you into a cloud of monatomic ions. These extreme stars also release vast amounts of energy in the form of flares, X-rays, and gamma-ray bursts, making them some of the most bizarre objects in the universe. One example of their extreme behavior was a flare in 2004 that compressed the magnetic field of the Earth from 50,000 light years away, despite a neutron star being only about 15 miles across but containing 1.5 times the mass of our sun.

336. **P**ulsars: The Intense Radiation Emitting, Speedy, and Precise Timekeepers. Pulsars are highly magnetized, rotating neutron stars that emit beams of electromagnetic radiation out of their magnetic poles, creating a lighthouse-like effect. These beams of radiation can be observed as regular pulses, which is where the name "pulsar" comes from. Pulsars are incredibly precise timekeepers, with some pulsars rotating hundreds of times per second, and they can be used for various astronomical measurements, such as detecting gravitational waves.

337. **P**SR J1748–2446ad: The Fastest-Spinning Pulsar. PSR J1748–2446ad, a rapidly rotating neutron star, has claimed the title of the fastest-spinning pulsar known to date, with an astonishing spin rate of 716 times per second, or 716 Hz. Discovered by astrophysicist Jason W. T. Hessels of McGill University on November 10, 2004, this pulsar has been the subject of much fascination and study since its confirmation on January 8, 2005.

338. **B**lack Widow Star Systems: When Pulsars Devour Their Companion Stars. Black widow star systems are binary star systems that comprise a pulsar, a dead star that spins rapidly, and a companion star of lower mass. The pulsar in these systems emits strong radiation that strikes its companion, gradually causing it to evaporate in a process called ablation. Like the spiders that share their name, the pulsars in black widow systems exploit their weaker companions by extracting material and energy from them. As the pulsar's radiation warms the companion star, it vaporizes and creates a disk of gas around the pulsar that is ultimately consumed.

339. **The Oldest Known Star: The Methuselah Star.** HD 140283, also known as the Methuselah star, is an astronomical object that challenges our understanding of the universe. According to scientific calculations, this star formed in the early universe and is estimated to be around 14.46 billion years old, which would make it older than the current age of the universe, estimated to be 13.79 billion years. This remarkable discovery has prompted scientists to revise their models of cosmic evolution. The Methuselah star, located in the constellation Libra, provides a unique window into the early history of the cosmos. It is one of the oldest known stars in the universe, and its age poses a conundrum for current theories of stellar evolution. Based on these theories, the star should have exhausted its fuel and exploded as a supernova long ago. However, it is essential to clarify that the age estimation for the Methuselah star is derived from scientific calculations rather than direct observation. Scientists are still trying to unravel the mystery of this ancient star and how it has managed to survive for so long.

340. **Betelgeuse: A Giant Star That Could Go Supernova Anytime.** Betelgeuse, located in the constellation Orion, is a red supergiant star that is over 1,000 times larger than our sun. Its sheer size and proximity to Earth make it one of the brightest and most recognizable stars in the night sky. But what sets Betelgeuse apart is the fact that it is nearing the end of its life and could go supernova at any moment. When it does, it will be one of the most spectacular and awe-inspiring events in the universe, with the explosion visible from Earth even during the daytime. Astronomers continue to study Betelgeuse to better understand the complex processes that govern the life and death of stars and prepare for the spectacular light show that could occur in the coming years.

341. **T**he Enigma of Tabby's Star: Unexplained Brightness Dips Puzzle Astronomers. Tabby's Star, also known as Boyajian's Star or KIC 8462852, is an intriguing star with an irregular and mysterious dimming behavior, exhibiting strange and unexplained dips in brightness, which has puzzled astronomers and generated significant interest from the public. These dips were first discovered by citizen scientists who were analyzing data from NASA's Kepler Space Telescope in 2011. The dips in brightness observed in Tabby's Star are unlike any other known star. They are sometimes deep and abrupt and can last for days or even weeks. In addition, they do not occur at regular intervals but appear irregular and unpredictable. Many hypotheses have been put forward to explain the strange behavior of Tabby's Star, including the presence of orbiting comets or a massive alien megastructure blocking the star's light, but subsequent studies have already ruled out this last explanation, so the star's behavior remains a puzzle. These enigmatic stars have baffled astronomers with their unprecedented peculiarities, leaving them scratching their heads in confusion.

342. **F**omalhaut: The Star with an Intriguing Debris Disk and a Controversial Exoplanet. Fomalhaut is a bright star about 25 light-years away from Earth. It is famous and controversial for its highly structured debris disk, which is the most symmetrical and elliptical ever observed around a star. The debris disk has a sharp inner edge, suggesting the presence of a massive planet orbiting the star. In 2008, a candidate planet, Fomalhaut b, was discovered within the disk using the Hubble

Space Telescope, making Fomalhaut one of the few stars with a confirmed exoplanet detected through direct imaging. However, subsequent observations raised doubts about the planet's existence, and its status remains uncertain.

343. **A** Star That Survived Its Own Eruption: The Enigmatic Resilience of Eta Carinae. Eta Carinae, a massive and luminous star located in the Milky Way galaxy, became one of the most prominent celestial objects in the sky after experiencing an extraordinary outburst in the 1840s, known as the Great Eruption. Despite emitting nearly as much light as a supernova explosion, the star survived and continued to fascinate scientists. Recent X-ray observations by NuSTAR have revealed that Eta Carinae is also one of the most potent sources of cosmic rays in our galaxy, posing a risk to astronauts and space electronics. The powerful eruption in the 19th century ejected material at unprecedented speeds, providing new insights into the complex behavior of stars and highlighting the importance of continued research in this field. These findings suggest that the blast was powered by a deep-seated explosion and not an extreme wind driven by the star's luminosity, doubling the kinetic energy of the eruption.

344. **V**ega: The Egg-Shaped Star with a Rotational Speed on the Edge of Destruction. Vega, located in the constellation Lyra, is the 5th brightest star in the night sky and looks very different up close. Unlike typical stars, Vega bulges into an egg-shaped oblate form due to its high rotational speed of over 600,000 miles per hour at the equator. This speed is estimated to be 90% of its maximum possible rotational speed, which, if exceeded by just 10%, would cause the star to be ripped apart due to the spin overwhelming Vega's gravity.

Furthermore, Vega's bulging equator shines less brightly than the poles, creating a dark stripe on its surface that scientists believe is caused by temperature variations resulting from the star's rapid rotation. The equator is cooler, which results in a darker color in contrast to the poles. In essence, Vega is like a blue Easter egg with stripes, and while it outshines nearby stars in the constellation Lyra, it hides its bizarre characteristics from the naked eye.

345. **P**ossible Star 'Cannibalism': A Star that Swallows a Neutron **Star.** A cosmic beast known as HV 2112, located 200,000 light-years away in the constellation Tucana, has been identified as a Thorne-Zytkow Object (TZO), a combination of a red giant and a neutron star that challenges our previous understanding of what constitutes a star. The neutron star was swallowed by its larger companion, a red supergiant star, resulting in a star that behaves differently than a conventional star and produces higher concentrations of heavy elements. Its discovery has forced scientists to rethink some of their most fundamental assumptions about the universe. Scientists identified HV 2112 by studying its unusual chemical composition, created by the interaction between the gas inside the supergiant and the hot surface of its inner neutron star. The discovery of HV 2112 has provided insight into possible star 'cannibalism,' as it is believed that the neighboring red supergiant star swallowed the neutron star to create this unique Thorne-Zytkow Object.

346. **T**he Vanishing Star: PSR J1841-0500, the Pulsar That **Occasionally Disappears.** Located about 22.8 light-years away from the Sun in the Scutum-Centaurus spiral arm of our galaxy, PSR J1841-0500 is a type of star known as a pulsar, which emits pulses of light as a result of its rapid spin. Its rotation period is about 0.9 seconds, which is considered typical for a pulsar star. However, what makes this star intriguing is that it tends to disappear from time to time. Initially believed to be a regular pulsar when it was discovered in December 2008, scientists studied PSR J1841-0500 for a year before it suddenly vanished just before they concluded their observations. Despite concerns that their equipment was malfunctioning, tests showed that the pulsar was indeed no longer present.

347. **Swift J1644+57: The Star That Vanished into a Black Hole.** Swift J1644+57 is a star located about 3.9 billion light-years away in the Draco constellation that was consumed by a black hole. The incident, which occurred in a smaller galaxy, was discovered when scientists detected an unusual amount of X-rays and gamma-rays from a previously inactive area of the universe. Upon further investigation, it was determined that the emissions came from a "jet" released after a black hole devoured a star. The jet was observed to be accelerating at a speed of 99.5% the speed of light away from the point of the event.

348. **The Ultra-Cool Stars: JO55-57AB and Their Extraordinary Properties.** The binary star system JO55-57AB, which consists of ultra-cool stars, has exceptional characteristics that challenge conventional classification methods, making it a fascinating object of study in the cosmos. Composed of two brown dwarfs, it has a surface temperature that is much cooler than our sun from 3,992 to 5,972 degrees Fahrenheit (2,200 to 3,300 degrees Celsius), while the surface temperature of the sun is about 9,932 degrees Fahrenheit (5,500 degrees Celsius) and emits most of its radiation in the infrared range. Despite its low luminosity, JO55-57AB exhibits a strong magnetic

field, which causes powerful flares and a constantly changing brightness. These features make it a prime target for scientists studying the dynamics and evolution of low-mass stars and brown dwarfs.

349. **The Flattest Star in the Sky: Achernar.** Achernar is a blue supergiant star located in the constellation Eridanus, about 139 light-years from Earth. It is the flattest star known in the sky, with its equator rotating at about 155 miles per second (250 kilometers per second), causing it to bulge at the equator and flatten at the poles. Achernar is also one of the brightest stars in the sky, with a luminosity about 5,000 times that of the Sun.

350. **The Super-Fast Star Ejected by a Black Hole: US 708.** One of the most famous hypervelocity stars is US 708, discovered in 2005. US 708 is moving at an astonishing speed of more than 745.645 miles per second (1,200 kilometers per second), making it one of the fastest-known stars in the Milky Way. Scientists believe that US 708 was once part of a binary system, but a black hole swallowed its companion star. The gravitational interaction with the black hole then flung US 708 away at such a high speed.

351. **The Most Luminous Stars and The Blazing Inferno of Wolf-Rayet Stars.** Wolf-Rayet stars are a phenomenon of the cosmos, capturing the imagination with their sheer size and luminosity. They emit vast amounts of ultraviolet radiation and strong stellar winds that strip away their outer layers, creating a nebula of gas and dust around them. These celestial behemoths are among the most massive stars known, with

surface temperatures that soar to 100,000 Kelvin, hot enough to ignite the air and create intense winds that whip around the stars at speeds of up to 6,710,000 miles per hour (3,000 km/s). This intense heat causes their atmospheres to be expelled, resulting in a rapid loss of mass over just a few million years. These stars are so important because they hold the key to understanding the evolution of massive stars and the production of heavy elements through stellar nucleosynthesis.

352. **The Star That Exploded Twice: The Puzzling. The behavior of FEN 121102.** FEN 121102 is a star that was observed to have exploded twice in a relatively short period of time. This is highly unusual as most stars only explode once in their lifetimes. The cause of FEN 121102's double explosion is not yet understood, and it has led to new theories about the nature of stars and their death.

353. **KIC 2856960: The Intriguing Behavior of a Triple Star System.** KIC 2856960 is a fascinating triple star system located in the Cygnus constellation, approximately 1,600 light-years from Earth. What makes this system special is the unusual behavior of its stars, with one of the stars exhibiting an extreme pulsating pattern never before seen in a triple-star system. The pulsating star, named KIC 2856960 A, is a Delta Scuti variable star that pulsates at two different frequencies, indicating that it is not a single star but a binary star system. The third star, KIC 2856960 C, orbits the binary system at a distance of about 2,800 astronomical units. The pulsating star in KIC 2856960 provides a unique opportunity for scientists to study the internal structure and evolution of multiple star systems. The cosmos has so much beauty, even in the strangest places.

354. **The Mesmerizing AR Scorpii: A White Dwarf Binary System Like No Other.** AR Scorpii (AR Sco) is a white dwarf binary system located approximately 380 light-years from Earth. It is a fascinating object in the sky that has captured the attention of astronomers worldwide. What makes AR Sco so special is its unique behavior that has never been seen before in any other white dwarf binary system. The system consists of a white dwarf, the remnant of a sun-like star, and a red dwarf companion. The two stars orbit each other every 3.56 hours, with the white dwarf being about twice the mass of the red dwarf. However, what makes AR Sco stand out is the emission of a powerful beam of radiation that pulses once every two minutes. This beam is caused by the interaction of the magnetic fields of the two stars, and it is so strong that it causes the entire system to brighten and dim in sync with the pulsations.

"

A hundred thousand million Stars make one Galaxy; A hundred thousand million Galaxies make one Universe. The figures may not be very trustworthy, but I think they give a correct impression.

– Arthur Eddington

SECTION 21
Galaxies: Exploring Cosmic Marvels and Mysterious Entities

355. **A** **Low Mass Galaxy with Surprising Properties: Segue 2.**
Segue 2 is a small, dim galaxy located about 114,000 light-years
away from Earth in the constellation Virgo. Despite its small
size, Segue 2 has a big mystery surrounding it. This tiny galaxy
has an extremely low mass and is composed almost entirely of
dark matter, making it one of the darkest matter-dominated
objects ever observed. In fact, Segue 2 has such a high dark
matter content that it challenges our understanding of
how galaxies form and evolve. Additionally, Segue 2 has an
unusually low number of stars for a galaxy of its size, further
adding to its mystery.

356. **T**he Tiny Titans of the Universe: The Presence of Ultra-
Compact Dwarf Galaxies. Ultra-compact dwarf galaxies
(UCDs) are small but dense objects that contain numerous
stars packed into a relatively small area. These galaxies are
believed to have formed from the merging of smaller galaxies
or from the tidal stripping of larger galaxies. Despite their
small size, UCDs contain large amounts of dark matter and
play a crucial role in the formation and evolution of galaxies.
Scientists are still studying UCDs to better comprehend their
properties and how they fit into the larger picture of the
universe's structure.

357. **T**he Cosmic Dwarfs in the Shadows: The Existence of T
Dwarfs. T dwarfs are a type of brown or failed star that is too
small to fuse hydrogen into helium and shine like a normal
star. Instead, T dwarfs cool and fade over time, becoming

redder and fainter. These objects are difficult to detect due to their low temperatures and faint luminosity, but their study is crucial for advancing our comprehension of the intricate processes involved in star formation and evolution.

358.   **T**horne-Zytkow Objects: The Rare and Fascinating Stellar Hybrids. Thorne-Zytkow Objects (TZOs) are some of the rarest and most fascinating objects. They are formed when a massive star collides with another star, creating a unique and unstable hybrid object. TZOs are characterized by a giant star with a neutron star at its core, surrounded by an incredibly dense and extended gas envelope. This envelope is so large that if a TZO were placed at the center of our solar system, it would extend beyond the orbit of Jupiter. One of the most exciting features of TZOs is their potential to host exotic phenomena, such as nuclear burning, that can occur in the outer envelope, producing heavy elements that are not typically formed in normal stars. This could lead to the discovery of new types of chemical compounds that are not found elsewhere in the universe.

359.   **E**lemental Factories: Thorne-Zytkow Objects' Potential to Create a Cosmic Chemistry Set. Thorne-Zytkow Objects are the only known objects in the universe where neutron capture can occur throughout the entire star, from the core to the outer layers. This means that they have the potential to produce a wide range of heavy elements, including those with atomic numbers greater than iron, which cannot be fabricated in significant amounts by normal stellar nucleosynthesis. Recent simulations suggest that Thorne-Zytkow Objects can create an abundance of heavy elements, such as barium, strontium, and yttrium, as well as rare earth elements. Some models predict

that TZOs could even produce elements like gold, platinum, and uranium. The exact extent of heavy element production in TZOs is still a matter of ongoing research, but the potential is exciting. With their ability to create a cosmic chemistry set, Thorne-Zytkow.

360. **Nebulae: The Cosmic Factories of Star Formation.** A nebula is a large cloud of gas and dust found in space. When a star dies, such as in a supernova, it can create a nebula by releasing its gas and dust. Other nebulae are areas where new stars are forming, which is why they are often referred to as "star nurseries." For stars to form in a nebula, clumps of dust and gas are pulled together by gravity. As these clumps grow larger, their gravitational force increases, causing the clump to eventually collapse under its own gravity. This results in a hot core at the center of the cloud, which marks the beginning of a star. Nebulae exist in interstellar space, the space between stars, and the closest known nebula to Earth is the Helix Nebula, which is believed to be the remains of a dying star similar to the Sun and located approximately 700 light-years away from us.

361. **The Boomerang Nebula: An Unprecedented Frigid Region in Space.** The Boomerang Nebula is an unparalleled celestial entity and one of the most frigid locations in the observable cosmos, with temperatures plummeting to a remarkable -521.6 degrees Fahrenheit (-272 degrees Celsius), or just 1 degree above absolute zero. It is a bipolar planetary nebula characterized by its double-lobed shape. The nebula is formed by the death of a star as it expels its outer layers, and the ejected material expands and cools, creating the characteristic cloud-like appearance of a nebula.

362. **The Red Rectangle Nebula: A Cosmic Chemistry Set for Studying Planetary Nebula Formation.** The Red Rectangle Nebula is a planetary nebula located about 2,300 light-years away from Earth. It is known for its distinctive red color and its X-shaped structure, which is thought to be caused by two binary stars at its center. The nebula also contains a unique mixture of chemicals, including carbon, oxygen, and nitrogen, which makes it a valuable laboratory for studying the processes that lead to the formation of planetary nebulae.

363. **The Pillars of Creation: A Cosmic Nursery for New Stars 7,000 light-years away.** A Cosmic Nursery, 7,000 Light-Years Away from Earth. The Pillars of Creation, comprised of three gas and dust towers located in the Eagle Nebula, have captured the public's imagination with their stunning appearance and symbolic representation of the universe's beauty. These formations serve as a vital location where new stars are born, allowing astronomers to gain valuable insights into the processes that shape the universe. Originally discovered by John Charles Duncan in 1920, the pillars were famously captured by the Hubble Space Telescope in 1995, revealing their rock-like appearance and complex interplay between gas, dust, and radiation. Moreover, the light we see from the Pillars is 7,000 years old, providing a glimpse into the past and allowing us to observe how they looked like thousands of years ago.

364. **The Crab Nebula: A Supernova in the Making.** The Crab Nebula, a celestial wonder located in the constellation Taurus, is a highly energetic and luminous supernova remnant formed from a star's explosion. This explosion was so powerful that it released an immense amount of energy, equivalent to the sun's output over 10,000 years, and created a shockwave that continues to expand at an incredible speed of over 3,728.23 miles per second (6,000 kilometers per second). Additionally, the Crab Nebula is home to a neutron star, one of the universe's densest objects, and rotates incredibly fast, completing a rotation in just 33 milliseconds! The Crab Nebula continues to amaze astronomers with its beauty and vast array of unique features.

365. **F**ormation of **Planetary Nebulae.** Planetary nebulae are expanding shells of glowing gas surrounding a dying star. As the star reaches the end of its life, it sheds its outer layers into space, leaving behind a small and hot core. The gas from the outer layers then ionizes due to the core's high temperature, creating the distinctive ring-like structure of a planetary nebula. These nebulae can be observed in a range of colors, from red to green, blue, and purple, depending on the elements present in the gas. Planetary nebulae significantly enrich the interstellar medium with heavy elements, providing the building blocks for future generations of stars and planets. In essence, planetary nebulae leave a lasting legacy that extends far beyond their ephemeral beauty.

366. **F**ormation of **Tidal Tails in Galaxies.** Tidal tails are elongated structures that can form when two galaxies interact and experience the mutual gravitational pull of each other. These tails can stretch out for hundreds of thousands of light-years and contain many stars, gas, and dust. Scientists believe that tidal tails play an important role in the evolution of galaxies, as the material within them can be funneled into the main galaxy, fueling further star formation and growth. The study

of tidal tails can provide valuable insights into the processes that drive galaxies' evolution and the universe's large-scale structure.

367. **T**he **Four Main Types of Galaxies: Spiral, Barred Spiral, Elliptical, and Irregular Galaxies.** Galaxies come in four main types: elliptical, irregular, spiral, and barred spiral, each of which can be further divided into subcategories. Spiral galaxies are the most common type, making up nearly 77% of all known galaxies, with the Andromeda galaxy serving as a prime example. Most spiral galaxies have a bar-like structure, and the Milky Way is no exception. Irregular galaxies are also prevalent, with approximately 25% of all known galaxies lacking a distinct shape. Elliptical galaxies, on the other hand, are mostly composed of old, low-mass stars and make up only around 10-15% of all known galaxies. Finally, the rarest type of galaxy is the elliptical doubled-ringed galaxy, also known as the Hoag-type galaxy, which comprises just 0.1% of all galaxies.

368. **T**he **Universe's Largest Objects: Fascinating Properties of Galaxy Clusters.** Clusters of galaxies are held together by gravity, with the largest ones containing thousands of galaxies spanning over ten million light years. These massive objects are ideal for studying the history of structure and galaxy formation, as well as the history of nucleosynthesis in the universe. Gravity in clusters mainly comes from dark matter, which makes them an excellent way to study it. Additionally, scientists study X-ray emissions from galaxy clusters, which come from hot gas trapped by the cluster's gravitational force, making up a larger part of the total mass of the cluster than the stars.

369. **S**tephan Quintet: A Cluster of Galaxies in Cosmic Dance. Stephan Quintet is a cluster of five galaxies, with four galaxies nearby interacting with each other. The interactions between the galaxies are intense, causing the gas in the galaxies to heat up to millions of degrees and produce bright X-ray emissions. The Stephan Quintet also contains a large amount of dark matter, estimated to account for more than 80% of the universe's mass. Despite the intense interactions, the galaxies are not expected to form a single, massive galaxy and will continue to move away from each other and interact with other galaxies in the surrounding space.

370. **B**rightest Galaxy in Sight: The Andromeda Galaxy. Even though several dozen smaller galaxies are closer to our Milky Way, Andromeda galaxy is the brightest external galaxy we can observe, excluding the Large and Small Magellanic Clouds, which may be seen from Earth's Southern Hemisphere. At 2.5 million light-years, it is the farthest object most humans can see without a telescope.

371. **T**he Most Distant and Earliest Galaxy Ever Observed: HD1. The potential galaxy known as HD1, estimated to be located about 13.5 billion light-years away from Earth, has the potential to be the farthest astronomical object ever observed. Its age places it in a time between the darkness of the universe's beginning, devoid of any stars or galaxies, and the emergence of light as clumps of gas and dust evolved into galaxies. The first galaxies formed roughly 100 million years after the Big Bang, much smaller and denser than the Milky Way, acting as building blocks in the construction of present-day galaxies like our own.

372. **Visible Galaxies from Earth: A Cosmic Window.** The galaxies that are visible from Earth with the naked eye are the Andromeda Galaxy, the Triangulum Galaxy, and the Milky Way, which is the galaxy that contains our solar system. The Magellanic Clouds, which are two small satellite galaxies of the Milky Way, are also visible. These galaxies can be seen under certain conditions, such as a clear night sky with minimal light pollution and a good viewing location. In addition, Omega Centauri and Sagittarius Dwarf Spheroidal Galaxy are visible from the Southern hemisphere or under very low light pollution conditions.

373. **The Dark Horse: Unusual Galaxy with Little to No Dark Matter**. Astronomers have discovered an unusual galaxy, named NGC 1052-DF2 or the "Dark Horse" galaxy, that appears to have no dark matter, which challenges current theories of galaxy formation. Dark matter is a mysterious, invisible substance that makes up about 27% of the universe, inferred from its gravitational influence on visible objects. The discovery was made by tracking stars in the outer regions of galaxies, which appear to be orbiting faster than their escape velocity, suggesting substantial mass holding them in orbit. NGC 1052-DF2 was identified using a low-budget setup of 48 commercial cameras and paparazzi-style lenses. This discovery opens a new frontier in astrophysics and deepens the mystery of dark matter.

374. **The Phantom Galaxy: The elusive and mysterious Galaxy X.** Galaxy X is a hypothetical large galaxy that is thought to be located in the outer reaches of the Milky Way. The existence of Galaxy X is inferred from the unusual orbits of small, distant objects in the Milky Way. While strong evidence suggests that

Galaxy X exists, it has yet to be directly observed. The search for Galaxy X has important implications for our understanding of the Milky Way and the potential for other undiscovered galaxies in the universe.

375. **Andromeda Galaxy: The Nearest Large Galaxy to the Milky Way.** The Andromeda Galaxy, also called M31 or NGC 224, is a spiral galaxy with a bar in its center, and a diameter of around 46.56 kiloparsecs, located approximately 2.5 million light-years away from Earth. It is the closest large galaxy to the Milky Way and was originally known as the Andromeda Nebula. Its name is derived from the constellation of Andromeda, which is named after the princess in Greek mythology who was married to Perseus.

376. **The Largest Known Spiral Galaxy in the Universe: NGC 687.** NGC 6872, the Condor Galaxy, is a massive barred spiral galaxy located in the constellation Pavo and is 212 million light-years away from Earth. It has two elongated arms and a diameter of over 522,000 light-years, 5 times the size of the Milky Way, making it the largest known spiral galaxy. NGC 6872 is interacting with a smaller lenticular galaxy called IC 4970. Discovered by John Herschel in 1835, it was always thought to be one of the largest stellar systems but was officially designated as the largest galaxy known to science by NASA in 2013. A new image of the galaxy combines visible, far-ultraviolet, and infrared light from different observatories, showcasing the galaxy's magnificence.

377. **One of the Rarest Things in Space: Ring Galaxies.** Only one in 10,000 galaxies belong to the rare and exceptional class of Ring Galaxies. The first of these, Hoag's Object, was discovered

in 1950. These galaxies boast a circular ring of bright stars surrounding a dark, empty center formed when two galaxies collide at a precise angle and velocity. The collision produces a shock wave that triggers star formation, resulting in a dense, spinning ring of stars in the galaxy's outer reaches. Despite the collision, the central regions of the galaxies pass through each other, leaving the center empty. These unique and intricate conditions make Ring Galaxies an uncommon sight in the cosmos, providing a fascinating opportunity for astronomers to study the complex and dynamic processes of the universe.

378. **The Formation of Polar-Ring Galaxies.** Polar-ring galaxies are a unique type of galaxy that have a ring of stars, gas, and dust orbiting perpendicular to the main plane of the galaxy. The most famous case of polar-ring galaxies is the Cartwheel galaxy, located 500 million light-years away from Earth. The formation of polar ring galaxies is a mystery that still intrigues astronomers today. One theory is that these galaxies are the result of a merger between two galaxies, with rings of stars and gas forming from debris from the smaller galaxy. Another theory proposes that polar ring galaxies form when a larger galaxy accumulates gas from a smaller satellite galaxy, which then stabilizes into a polar orbit around the main galaxy.

379. **The Galaxy That Time Forgot: The Unique Characteristics of DGSAT.** DGSAT, or Dragonfly 44, is a galaxy dubbed "The Galaxy That Time Forgot." It is a relatively new discovery, having only been observed in 2015, and is located about 300 million light-years away in the Coma Cluster. What makes DGSAT unique is its composition: it is made up almost entirely of dark matter, with very few stars. It is estimated that only about 0.01% of its mass consists of stars. This has led to speculation that DGSAT is not a galaxy, but rather a mass of dark matter mistaken for a galaxy.

380. **T**he **Whirlpool Galaxy: A Spiral Beauty.** M51, commonly referred to as the Whirlpool Galaxy, is a grand-design spiral galaxy renowned for its striking, sweeping arms composed of long lanes of stars, gas, and dust. These arms serve as cosmic "nurseries," stimulating the formation of new stars by compressing hydrogen gas to generate star clusters. While the exceptional prominence of the Whirlpool's arms has puzzled astronomers for some time, some theories suggest that a close encounter with the small, yellowish galaxy NGC 5195 at the outermost edge of one arm may be responsible for this captivating feature. Notably, the Whirlpool Galaxy was the first object to be identified as a spiral galaxy, earning its place as a cherished object of study and admiration among astronomers and stargazers alike. However, its distinctive helical shape is not solely a product of its grand spiral structure. The gravitational influence of its central supermassive black hole plays a vital role in shaping the galaxy's overall appearance, affecting the movements and alignments of stars and other celestial bodies. Consequently, the Whirlpool Galaxy's spiral pattern emerges from a complex interplay of astrophysical processes, resulting in one of the most stunning and awe-inspiring sights in the universe.

381. **T**he **Stellar Showdown: The Whirlpool Galaxy.** The Whirlpool Galaxy, also known as M51 or NGC 5194, is a breathtaking object in the night sky engaged in a cosmic dance with its smaller companion galaxy, NGC 5195. As they spiral towards each other, their gravitational interaction triggers bursts of star formation and creates bright blue knots of stars along the galaxy's arms, shaping the evolution of both galaxies. This stunning display showcases the boundless and mutable nature of the cosmos.

382. **G**alactic Cannibalism: Larger Galaxies Absorb Smaller Ones. Galactic cannibalism is a phenomenon in which a large galaxy absorbs a smaller one by using its gravitational force, stretching and tearing it apart. The stars, gas, and dust from the smaller galaxy merge with those of the larger one, changing its shape and composition. This process is frequently observed in dense clusters, where galaxy collisions occur frequently, and gas-poor spiral systems and cD galaxies with multiple nuclei can be found. These effects are the result of violent interactions and gravitational fields that enable massive central galactic systems to capture smaller cluster members, effectively engaging in galactic cannibalism.

383. **T**he Powerhouses of Galaxies: The Astonishing Active Galactic Nuclei. Active galactic nuclei are now understood to be active supermassive black holes at the centers of galaxies that emit jets and winds, shaping their host galaxies. These emissions of gas and dust are detected by telescopes at various wavelengths along the electromagnetic spectrum, from X-rays and visible light to infrared and radio waves. Quiescent black holes, in contrast, do not emit any detectable light.

384. **H**oag's Object: A Bizarre Perfect Ring Galaxy in Space. In 1950, astronomer Arthur Hoag discovered an unusual extragalactic object that raised the question of whether it was one galaxy or two. This object features a ring dominated by bright blue stars on the outside, with a ball of much older, redder stars near the center. The gap between them appears nearly dark. The origin of Hoag's Object is still a mystery, but possible explanations include a galaxy collision or the gravitational effect of a central bar that no longer exists. Recent observations show that Hoag's Object has not accreted a smaller galaxy in the past billion years. The object is located about 600 million light-years away in the Serpens constellation and spans about 100,000 light years. The Hubble Space Telescope recently reprocessed a photo of it using an AI de-noising algorithm. Visible in the gap is another ring galaxy, while many distant galaxies are visible in the background.

385. **V**ast **Dark Spaces: Cosmic Voids.** These "dark spaces" contain a mix of galaxy clusters and matter, unlike the rest of the universe. These voids are large regions of the universe with very few or no galaxies. Compared to the entire universe, the cosmological evolution of the empty regions deviates significantly: there is a long stage when the curvature term predominates, which prevents the formation of clusters and massive galaxies. Therefore, even though even the most barren areas of voids contain more than 15% of the universe's normal matter density, the voids appear to be nearly empty to an observer. They can have diameters of up to 150 million light-years and can make up to 80% of the universe's volume. Maybe the void is trying to tell us something - that sometimes, less is more.

386. **S**upernova **Remnants Shape Our Galaxy: Types and Classification.** Supernova remnants (SNRs) are the aftermath of a supernova explosion, and they play a significant role in our understanding of the galaxy. They heat the interstellar medium, distribute heavy elements, and accelerate cosmic

rays. SNRs are classified into three types: shell-type, crab-like, and composite remnants. Shell-type remnants appear as a ring-like structure and are heated by the shockwave from the supernova. Crab-like remnants are filled with high-energy electrons that interact with the magnetic field and emit X-rays, visible light, and radio waves. Composite remnants are a combination of both shell-like and crab-like and can be further categorized into thermal and plerionic composites based on their X-ray spectra. By observing SNRs, astronomers can better understand the evolution of stars and the universe.

387. Cataclysmic Variables: A Look into the Mysterious Binary Star Systems. Cataclysmic Variables (CVs) are small yet fascinating binary star systems consisting of a white dwarf and a normal star companion. Their size is similar to that of the Earth-Moon system, and they have an orbital period of 1 to 10 hours. As the companion star loses material to the white dwarf through accretion, the enormous gravitational potential energy is converted into X-rays. While there are over a million of these CVs in our galaxy, only a few hundred have been studied in X-rays as they are faint in comparison to other X-ray sources. Delve into the world of CVs and explore the intriguing mysteries of these binary star systems.

"

The planets, to whom no place is barren or unattractive, however devoid of charm it may be to mankind, are the sources of manifold good and useful things.

– Johannes Kepler

SECTION 22
Cosmic Collisions: Explosive Encounters that Shape the Universe

388. **Galactic Mergers: The Cocooned Black Holes That Devour Everything.** Black holes are often portrayed as all-consuming entities, but they can be surrounded by orbiting stars, gas, and dust for long periods of time. When two galaxies merge and their central black holes approach each other, the gas and dust in the vicinity are pushed toward their respective black holes, creating what astronomers call an active galactic nucleus (AGN). The AGN becomes enshrouded in a cocoon of gas and dust as the black holes consume material spiraling rapidly toward them, emitting high-energy radiation. A recent study using NASA's NuSTAR telescope found that the AGN is heavily obscured during the late stages of a merger, confirming the longstanding theory that the black hole does most of its eating during this time. By comparing NuSTAR observations with data from other observatories, researchers were able to determine which galaxies had heavily obscured AGNs. The study helps increase our understanding of the relationship between a black hole and its host galaxy.

389. **Shoemaker-Levy 9: A Celestial Collision of Unprecedented Magnitude.** In 1994, the comet Shoemaker-Levy 9 collided with Jupiter, leaving a scar on the gas giant larger than Earth. The comet fragmented under Jupiter's gravitational pull, resulting in a series of impacts over several days. Each collision unleashed powerful explosions, leaving dark scars visible from Earth. The largest scar, called "G," measured an astounding 12,000 kilometers (7,500 miles) in diameter. This extraordinary event provided invaluable insights into planetary impacts and the composition of comets. It serves as a striking reminder of the immense forces shaping our cosmic neighborhood.

390. **C**rash Landing: The Moon's Giant Impact Hypothesis. The Giant Impact theory has emerged as a prominent contender to explain the origin of the Moon, suggesting that during Earth's early formation stage, it collided with a neighboring emerging planet called Theia. This catastrophic event resulted in the temporary breakup of both planets into gaseous, molten, and elemental globules, which later reformed into the familiar Earth and Moon. Recent research provides additional evidence that strengthens the validity of this theory.

391. **C**helyabinsk: The Explosive Impact of a Mid-Sky Asteroid on Earth In 2013, a small asteroid measuring approximately 65 feet (20 meters) in diameter entered Earth's atmosphere, hurtling at a speed of roughly 12 miles per second (~19 km/sec). The asteroid collided with Earth's protective atmospheric layer, causing it to detonate at an altitude of approximately 20 miles (30 kilometers) above the Russian city of Chelyabinsk. The resulting explosion was incredibly bright and hot and had a force equivalent to 20 to 30 times the energy released by the atomic bomb that was dropped on Hiroshima during World War II. The shockwave generated by the explosion shattered windows and caused damage to buildings in six Russian cities, leading to injuries for approximately 1,500 people, most of whom were harmed by flying shards of glass.

392. **T**he Great Saturnian Ring Collision: Saturn's Rings Created by Collision. In the early 1980s, a collision occurred between two of Saturn's rings, resulting in a bright flash of light and the

formation of a new ring. This collision has helped researchers understand how Saturn's rings may have formed. If colliding moonlets were made up of only small, icy particles, they would have disintegrated, resulting in only a ring. However, if the moonlets had denser cores, they might not have completely destroyed each other, resulting in two remnant moons with half the mass of the originals. NASA's Cassini spacecraft found that both Prometheus and Pandora had dense cores, matching this theory. This provides insight into the dynamic nature of the universe.

393. **The Great White Dwarf Smash: Type 1a Supernovae.** A Type 1a supernova occurs when a white dwarf star, the remnant of a small star like the Sun, steals matter from a companion star until it reaches a critical mass and explodes. These highly bright explosions are used as "standard candles" to measure the distance to faraway galaxies. The intense brightness and uniformity of Type 1a supernovae make them ideal for measuring astronomical distances. By studying the light from these supernovae, astronomers can estimate their distance from Earth, helping them map the universe'. This phenomenon, nicknamed the "Great White Dwarf Smash," is an essential tool for understanding the nature of the universe. Sometimes, in the grand celestial dance of the stars, two partners collide with such force that the very fabric of space-time trembles.

394. **An Expected Cosmic Collision: The Milky Way's Destiny with Andromeda.** In approximately 4 billion years, the Milky Way and Andromeda, two of the largest galaxies in our local group, are predicted to collide, resulting in an awe-inspiring cosmic event. While the central regions of the two galaxies

are expected to avoid a direct collision, the outer halos will interact gravitationally, creating a mesmerizing cosmic dance. While there are various theories regarding the outcome of this collision, it is challenging to predict accurately. Some researchers have suggested that the two galaxies will merge to form a new, larger galaxy, potentially an elliptical galaxy, while others propose that it could result in an explosion, or that the two galaxies will pass through each other mostly unchanged. However, regardless of the outcome, the Milky Way's expected collision with Andromeda will be a momentous event in the universe's history and will have a significant impact on the formation of galaxies for billions of years to come.

395. **T**he Bullet Cluster: Dark Matter Caught in the Act of Separating from Visible Matter in Cosmic Collision. The Bullet Cluster, also called 1E 0657-56, is a remarkable astronomical object that has drawn attention from both scientists and the public. It is a galaxy cluster located approximately 3.8 billion light-years away in the constellation Carina. What sets the Bullet Cluster apart is that it provides evidence for the existence of dark matter, an enigmatic substance that cannot be observed directly but can be inferred from its gravitational effects on visible matter. In 2006, observations of the Bullet Cluster revealed that there were two clusters in the process of colliding. As the clusters collided, their gas interacted and slowed down while the dark matter continued unimpeded, resulting in the visible matter lagging behind the dark matter. This phenomenon, known as gravitational lensing, provided powerful evidence for dark matter's existence and its separation from visible matter.

396. **M**ilky Way's Collision Preview: Antennae Galaxies Create Super Star Clusters. A Fertile Marriage of When galaxies collide, it can cause bursts of star formation and reshape their appearance. The Antennae Galaxies are a well-known example, located 45 million light-years away from Earth. The collision created a complex system of tidal tails, bridges, and young blue star clusters. Hubble Space Telescope observations revealed that billions of stars were born from the interaction of the two spirals, with about 10% forming super star clusters that will likely survive to become globular clusters. The Antennae Galaxies offer a glimpse of what may happen when our Milky Way collides with Andromeda in about 6 billion years.

397. **T**he Hercules-Corona Borealis Great Wall: Universe's Largest Structure Challenges Inflation Laws. The largest known structure in the universe is the Hercules-Corona Borealis Great Wall, which is a supercluster so massive that it challenges the laws of inflation. In the 17th century, Ole Roemer discovered that the speed of light was not infinite, but rather finite. After years of precise measurements, scientists determined the speed of light to be an astounding 299,792,458 meters per second. This means that even light from the closest star takes 4.24 years to reach Earth, and a standardized unit for measuring astronomical distances, the light-year, was established. The Milky Way galaxy is 100,000 light-years wide, but the Hercules-Corona Borealis Great Wall dwarfs it in size, making it almost impossible to comprehend.

398. **The Tadpole Galaxy: Trailing Starry Debris from a Collision.** The galaxy UGC 10214, also known as the "Tadpole," is an unusual spiral galaxy that appears to be moving quickly through space. Its shape was distorted by a smaller blue galaxy, visible in the upper left corner, that collided with it and left a long trail of stars and gas debris stretching more than 280,000 light-years. The collision also created numerous young blue stars and star clusters in the spiral arms and the long "tidal" tail of stars, which will eventually become old globular clusters. Two clumps of stars in the tail are likely to become dwarf galaxies orbiting in the Tadpole's halo. The galaxy collision occurred against the backdrop of a wallpaper pattern of 6,000 galaxies, which were captured in the same image as the Tadpole by the Advanced Camera for Surveys aboard the Hubble Space Telescope. This picture took one-twelfth of the time to capture compared to the Hubble Deep Field and reveals even fainter objects. These galaxies represent a fossil record of the universe's evolution over 13 billion years.

399. **T**he **Cartwheel Galaxy: Shockwave Loops and Ultraviolet Brightness Amaze Astronomers.** This galactic structure originated from a high-velocity impact that took place approximately four hundred million years ago. The Cartwheel is made up of a pair of loops, a luminous interior loop, and a vivid exterior loop, that emanate outwards from the epicenter of the crash in a manner similar to shockwaves. In addition to its unique structure, the Cartwheel Galaxy is known for its remarkable characteristics. The outer ring of the galaxy rotates at an impressive speed of 217 kilometers per second, while simultaneously appearing to move away from us at an astonishing rate of 9,050 kilometers per second. Scientists estimate that the Cartwheel Galaxy's mass is somewhere between 2.9 and 4.8 billion times that of our Sun. Furthermore, this galaxy is a significant source of ultraviolet radiation, making it one of the brightest such sources in the nearby cosmos.

400. Gamma-ray Bursts: The Brightest Cosmic Explosions Created by Black Hole Formation. Gamma-ray bursts (GRBs) are extremely bright and short-lived bursts of gamma-ray light, that shine hundreds of times brighter than a typical supernova and about a million trillion times as bright as the Sun. They were discovered in the late 1960s by U.S. military satellites, and until recently, they were the biggest mystery in high-energy astronomy. The energy behind a gamma-ray burst is believed to come from the collapse of matter into a black hole. There are two types of GRBs, long-duration, and short-duration. Long-duration bursts are associated with the deaths of massive stars in supernovas, while short-duration bursts appear to be associated with the merger of two neutron stars into a new black hole or a neutron star with a black hole to form a larger black hole. Recent observations from satellites like Swift and Fermi have provided evidence for these theories.

"

*Every one of us is, in the
cosmic perspective, precious.
If a human disagrees
with you, let him live. In a
hundred billion galaxies, you
will not find another.*

– Carl Sagan

SECTION 23
Mysteries of the Cosmos: Strange and Enigmatic Phenomena

401. **The Fermi Bubbles: A Groundbreaking Discovery of Massive Gamma-Ray Structures in the Milky Way.** Fermi bubbles are enormous regions that emit gamma rays, extending over 50,000 light-years on both sides of the Milky Way's center. These bubbles protrude outside the galaxy's plane, and despite their size, their origin has been a mystery. Scientists have proposed different hypotheses for their formation, including black hole activity, winds from the black hole, and continuous star formation. A recent study utilized X-ray observations from a satellite to investigate the formation of bubbles in space. The simulations showed that winds from a black hole traveling at high speeds over a long period, inject energy into the surrounding gas. These winds are streams of charged particles that interact with nearby gas, causing a reverse shock that creates a distinct temperature peak. The inside volume of this reverse shock front corresponds to the observed bubbles. The study suggests that explosive activity at the center of the galaxy could not explain the formation of the bubbles, and instead, the winds from the black hole are the likely cause.

402. **The Fermi Bubbles and the Big Bang: A Surprising Cosmic Connection.** One fascinating discovery about the Fermi Bubbles is their impact on the cosmic microwave background radiation, which is the afterglow of the Big Bang. The bubbles appear to be blocking some of the radiation in certain areas, creating a "cold spot" in the cosmic microwave background map. Some astronomers have jokingly referred to the cold spot as "Elsie's Nipple," after a prize cow with a similar feature on its udder. While it's crucial to take scientific discoveries seriously, a little levity can help lighten the mood when exploring the mysteries of the universe.

403. **Irregular Rotation of Celestial Bodies: A Puzzling Phenomenon.** Celestial bodies, including planets and stars, are not always expected to rotate in a regular, symmetrical manner. Irregular rotation is a common phenomenon in celestial bodies and can provide valuable insights into their internal structure and formation. For example, the irregular rotation of Saturn's moon Iapetus suggests that it has been impacted by another celestial body in the past. Similarly, the rapid and irregular rotation of some asteroids they were once part of larger, shattered bodies. Understanding the causes of irregular rotation can also help us better understand the evolution of celestial bodies and the impact of collisions and other events in shaping their orbits.

404. **Blazars: The Cosmic Giants that Shape the Universe.** Blazars are some of the most energetic and powerful objects in the universe; they can produce jets of particles and radiation that extend across millions of light-years. These jets are powered by the supermassive black holes at the centers of blazar galaxies, which can have masses billions of times greater than the Sun. As material is pulled into the black hole, it heats up and emits intense radiation, which drives the production of the high-speed jets. The energy of these jets can exceed the power output of an entire galaxy, and they can shape the universe's structure on vast scales. Some scientists believe that blazars may play a key role in regulating the growth of galaxies and the distribution of matter in the cosmos.

405. **B**lazars: **The Extreme AGN Jets That Could Be Cosmic Neutrino Sources.** Blazars are a type of active galactic nuclei (AGN) that have jets pointed toward the observer, making them the most extreme subclass of AGN. These jets emit high-energy photons, which dominate the extragalactic gamma-ray sky and reach multi-TeV energies. This suggests that blazars can accelerate particles to very high energies, including protons that could potentially make them cosmic neutrino sources. Blazars are rare and bright, making them ideal candidates for testing this theory. Recently, several multi-messenger monitoring campaigns have been launched in response to high-energy neutrinos detected by the IceCube Neutrino Observatory from the direction of blazars. This article summarizes the theoretical interpretation of these observations and provides an overview of the potential role of blazars as neutrino sources based on experimental results.

406. **T**he Mysterious Hole in NGC 247: An Unexplained Phenomenon. NGC 247 is a stunning spiral galaxy, mostly seen from its edge. Despite being smaller than the Milky Way, it contains some bright nurseries where new stars are born. However, its most striking feature is the dark void on one side of its core, which looks as though a hole has been punched through the disk. While some astronomers initially attributed this void to gravitational interactions with a nearby galaxy, more recent research suggests that a blob of dark matter could have plunged through the disk, scattering stars, and blowing away gas and dust. Alternatively, some have proposed that the void might be the result of extraterrestrial intelligence constructing Dyson Spheres around nearby stars, starting with their own star and expanding outwards, resulting in a hole in the galaxy that grows as they expand. While the hole in

NGC 247 is probably natural, the possibility of extraterrestrial involvement cannot be entirely ruled out. It's a fascinating case that raises many questions, such as what their technology would be like and how they would power it.

407. **T**he Behavior of Matter in Accretion Disks of Active Galactic Nuclei. An active galactic nucleus (AGN) is a compact region at the center of a galaxy that is believed to contain a supermassive black hole. The intense gravitational field of the black hole causes matter in the surrounding accretion disk to rapidly accrete, releasing a large amount of energy in the process. This energy is considered to drive the powerful emission from the AGN, which can be seen across the entire electromagnetic spectrum. The behavior of matter in these accretion disks is a subject of ongoing study, with scientists attempting to understand the complex physical processes that govern the flow and emission of matter in these extreme environments.

408. **T**he Variability of Luminosity in Active Galactic Nuclei. Active galactic nuclei (AGN) are some of the most luminous objects in the universe, and their luminosity can vary significantly over time. This variability is thought to be caused by several factors, including changes in the accretion rate onto the central supermassive black hole, variations in the jet emission, and instabilities in the accretion disk. The study of the variability of AGN luminosity is important for understanding the physical processes that govern the emission from these objects and for developing methods for detecting and characterizing AGN in large-scale astronomical surveys.

409. **The Existence of Ultra-Luminous X-Ray Sources.** Ultra-luminous X-ray sources (ULXs) are a class of celestial objects that emit X-rays with luminosities that are far greater than what would be expected from a single star. The origin of these objects is not well understood, but they are thought to be powered by intermediate-mass black holes, accreting material from a companion star. The study of ULXs is crucial for unlocking the secrets of these enigmatic black holes and understanding the extreme environments that give rise to their bright, flashy performances.

410. **The Great Attractor: A Massive Gravitational Anomaly Pulling Galaxies Towards It.** The Great Attractor, located around 150 Mpc (approximately 3.26 light-years or 3.086×10^{13} kilometers) away from Earth in the direction of the Centaurus and Norma clusters, is a massive gravitational anomaly with an estimated mass of around 10^{17} solar masses. It acts like a cosmic vacuum cleaner, pulling not just our galaxy but also many others toward it. This massive gravitational pull is responsible for the motion of not just the Milky Way but also other nearby galaxies. Despite its immense size and influence, the Great Attractor remains a mystery to scientists who are working tirelessly to unravel its secrets and better understand its significance in the cosmos.

411. **Cosmic Illusions: A Stunning Phenomena Created by Ice Crystals.** Cosmic illusions are fascinating and beautiful atmospheric phenomena that occur when light interacts with ice crystals in the atmosphere. These illusions include halos, pillars, arcs, and circles that can be seen in the sky, often during sunrise or sunset. They result from the refraction and reflection of light by the hexagonal-shaped ice crystals, which act as prisms, bending and reflecting the light to create colorful circular or semi-circular shapes. Cosmic illusions not only have aesthetic appeal but also hold significant scientific value. They offer insights into various atmospheric elements, such as ice crystal size and shape, atmospheric composition and structure, and light distribution under specific atmospheric conditions. They also aid in understanding precipitation formation and weather pattern development. For instance, the 22-degree halo, the most frequent halo type, can determine the height and thickness of high-altitude cirrus clouds. Similarly, the circumzenithal arc, appearing as an inverted rainbow, is generated by sunlight passing through ice crystals at a specific angle, providing information about the ice crystals' size and shape.

"

A book, too, can be a star, a living fire to lighten the darkness, leading out into the expanding universe.

– Madeleine L'Engle

SECTION 24
Theories, Paradoxes, and Puzzles

412. **E**instein's **Theory: Relativistic Speeds and Observed Frequencies.** When objects are moving at speeds close to the speed of light, the length observed by the moving observer will contract, and time will dilate. This phenomenon affects the frequency of the waves observed at relativistic speeds. The Lorentz transformations are used to calculate the geometrical coordinates and time from one space system to another moving at a constant velocity. The relativistic Doppler effect equation can be used to calculate the observed frequency when taking these effects into account. Additionally, the frequency observed by an observer moving closer to a star can be calculated using classical and relativistic equations. Redshift occurs when the source of light moves away from the observer.

413. **T**he **Faint Young Sun Paradox: The Mystery of Earth's Liquid Water Despite the Faint Young Sun.** The Faint Young Sun Paradox, also known as The Early Sun's Enigma, has perplexed scientists for many years. The paradox refers to the issue of Earth having liquid surface water despite the Sun's lower luminosity during its early history, which should have caused the Earth to freeze due to the lower temperature and brightness of the Sun. Scientists have proposed several theories to resolve this paradox, including the presence of high concentrations of greenhouse gases, particularly methane and carbon dioxide. These gases caused Earth to warm up to life-sustaining levels. As solar radiation increased, greenhouse gas concentrations must have decreased proportionally to prevent overheating. Rock weathering and biological processes are believed to have regulated greenhouse gas concentrations during this time.

414. **P**lanck's Law: A Fundamental Breakthrough in Understanding Black-Body Radiation. Planck's law is a physics theory that explains the spectral density of electromagnetic radiation emitted by a black body at a given temperature T, assuming that there is no net flow of matter or energy between the body and its environment. In the late 19th century, scientists were puzzled by the fact that the observed spectrum of black-body radiation diverged significantly at higher frequencies from that predicted by existing theories. In 1900, Max Planck devised a formula for the observed spectrum by hypothesizing that the energy of a hypothetical electrically charged oscillator in a cavity containing black-body radiation could only change in a minimal increment, E, that was proportional to the frequency of its associated electromagnetic wave. Although Planck initially thought this hypothesis was a mathematical trick, other physicists such as Albert Einstein further developed his work, and today, Planck's insight is widely recognized as a fundamental contribution to quantum theory.

415. **Q**uantum Gravity Theories: The Quest to Unify the Fourth Fundamental Force. The standard model of particle physics covers three fundamental forces, but quantum gravity, the fourth force, has not been successfully quantized due to the difficulty of testing proposals on extremely small scales. Canonical quantum gravity, loop theory, and string theory are the most important versions of quantum gravity theories. While the other three forces act in space-time, gravitation is identified with the curvature of space-time itself, making it more difficult to deal with. Some propose that space-time should be deprived of its fundamental status and shown to emerge in a non-spatio-temporal theory to address this challenge.

416. **S**tring Theory: Exploring the Fundamental Objects of the Universe Beyond Particles. String theory is a scientific theory in the field of particle physics that aims to merge quantum mechanics and Albert Einstein's general theory of relativity. String theory proposes strings as the fundamental objects of the universe, in contrast to particles. These strings can vibrate in different ways, each vibration corresponding to a particle with specific characteristics like mass and charge. During the 1980s, physicists recognized that string theory held the potential to unify the four fundamental forces of nature—gravity, electromagnetism, strong force, and weak force—along with all forms of matter, within a single framework based on quantum mechanics. This suggested that it could be the sought-after unified field theory. However, despite ongoing rapid development and active research, string theory has yet to establish direct experimental confirmation, remaining primarily as a mathematical construct.

417. **C**yclic Universe Model: Infinite Self-Sustaining Expansion and Contraction. A cyclic model, also known as an oscillating model, is a cosmological theory that suggests the universe goes through infinite or indefinite self-sustaining cycles. One example is the oscillating universe theory, which was briefly explored by Albert Einstein in 1930. This theory proposes that the universe experiences a continuous series of oscillations, with each cycle starting with a Big Bang and ending with a Big Crunch. During the cycle, the universe would expand for

a period before gravity eventually caused it to contract and undergo a bounce.

418. **Inflation Theory: How a Rapid Expansion of the Universe Solved Big Bang Puzzles.** The Inflation Theory is a scientific concept developed in 1980 to explain the puzzles in the Big Bang Theory. It combines ideas from quantum and particle physics to explore the universe's first moments following the Big Bang. This theory proposes a period of rapid expansion, known as inflation, that occurred in the universe's early stages. This event led to several consequences, such as the universe's size being much larger than expected, and the uniform distribution of energy and flat geometry of spacetime that the Big Bang Theory could not explain. The Inflation Theory was introduced by particle physicist Alan Guth and is now widely accepted as a crucial component of the Big Bang Theory.

419. **The Steady-State Theory: The Universe as an Eternal and Unchanging Entity.** The steady-state theory, proposed by Sir Hermann Bondi, Thomas Gold, and Sir Fred Hoyle in 1948, suggests that the universe is eternal and unchanging. As new matter is created to fill the void left by the universe's expansion, the average density of matter remains constant. The universe has always existed and will always exist in a steady state.

420. **The Many-Worlds Interpretation: Exploring the Possibility of Infinite Parallel Universes.** The many-worlds interpretation, proposed by physicist Hugh Everett in 1957, suggests that there are infinite parallel universes, each with its own version of reality. Whenever a quantum event occurs, the universe splits into multiple universes, each with a different outcome.

Some scientists believe that the multiverse could explain why our universe seems fine-tuned for life and why there are so many coincidences and patterns.

421. **The Holographic Principle: Is the Universe a 2D Projection?** Proposed by physicist Gerard 't Hooft in 1993 and expanded upon by physicist Leonard Susskind in 1995, the Holographic Principle suggests that the information in a region of space can be encoded on its boundary. In simpler terms, it means that the universe's information can be contained in a 2D surface, much like a 3D image can be encoded on a 2D hologram. According to this theory, the universe we perceive as 3D is actually a projection of a 2D reality. In other words, everything we see and experience in the universe may be a holographic projection from a distant two-dimensional realm. This concept has been explored in various areas of physics and has potential implications for understanding the nature of reality.

422. **The Ekpyrotic Universe: A Cycle of Expansion and Contraction.** The Ekpyrotic Universe, proposed by Paul Steinhardt and Neil Turok in 2001, suggests that our universe is an infinite cycle of expansion and contraction resulting from the collision of two branes in a higher-dimensional space. This collision is known as the "ekpyrotic scenario," leading to a Big Bang-like event on the branes, causing the universe to expand rapidly.

423. **The Chaotic Inflationary Universe: Self-Creating and Eternal** The chaotic inflationary universe theory, proposed by physicist Andrei Linde in 1983, suggests that the universe is self-creating and eternal. The universe goes through an infinite cycle of inflation and expansion, giving rise to new universes with each inflationary period.

424. **T**he Fermi Paradox: The Mystery of the Missing Aliens. The Fermi paradox is an apparent contradiction between the probable existence of extraterrestrial civilizations and the lack of evidence or contact with such civilizations. The Fermi Paradox raises the question of why, despite the high likelihood of the existence of extraterrestrial intelligence, we have not found any evidence of it. The paradox was named after Enrico Fermi, a Nobel Prize-winning physicist who used back-of-the-envelope calculations to estimate the probability of extraterrestrial life. The paradox suggests that, given the vast number of stars and planets in the Milky Way, intelligent life should exist on some fraction of Earth-like planets. However, despite the possibility of advanced technology and interstellar travel, we have not yet made contact with any extraterrestrial beings. The Fermi Paradox is a subject of ongoing debate and has led to various proposed solutions.

425. **T**he Wave-Particle Paradox: The Mystery of the Double-Slit Experiment. Wave-particle duality is a concept in quantum mechanics that states every particle or quantum entity can be described as either a particle or a wave, and the classical concepts of "particle" and "wave" are insufficient to fully explain their behavior. This duality has been confirmed not only for elementary particles but also for compound particles such as atoms and molecules. While wave-particle duality has been useful in physics, its interpretation has not been fully resolved. Niels Bohr saw this duality as a fundamental fact of nature, and Werner Heisenberg viewed it in the context of quantum field theory. Heisenberg proposed that duality exists for all quantum entities and generated a new concept of fields, which replaced classical field values. Ordinary quantum mechanics is then seen as a specialized consequence of quantum field theory.

426. **The Drake Equation: The Probability of Alien Life.** Have you ever wondered if there are other intelligent life forms out there in the universe? The Drake Equation is a method for estimating the number of extraterrestrial civilizations in the Milky Way galaxy that might be capable of communicating with Earth. Scientists have been pondering this question for a long time, and one of the ways they try to estimate the possibility is through something called the Drake Equation. It was named after Frank Drake, an astrophysicist who came up with the equation in 1961. Essentially, it's a mathematical formula that tries to estimate the number of extraterrestrial civilizations in the Milky Way galaxy that could communicate with Earth.

427. **The Anthropic Principle: The Coincidence of Life.** The Anthropic Principle is a controversial idea that suggests the universe is finely tuned for the existence of life. This principle asserts that the laws of physics and the universe's initial conditions are precisely calibrated so that life, as we know it, can exist. If these parameters were even slightly different, the emergence of sentient life would not have been possible. Some scientists believe this fine-tuning is evidence of intelligent design, although this notion is disputed. Despite its controversial nature, the Anthropic Principle emphasizes the remarkable coincidence of life in the universe. It proposes that conscious observers are not merely coincidental but rather

428. **The Universe's Whisper: The Unsolved Mystery of the Mysterious Radio Signs.** Scientists have detected a mysterious radio signal from deep space that has yet to be explained. The signal, known as the "fast radio burst," lasts only a few milliseconds and has been detected a handful of times. Some theories suggest that the signal could come from a neutron star or even an extraterrestrial civilization, but no concrete evidence has been found to support these claims. Despite ongoing research, the source of the mysterious radio signal remains a mystery. As the renowned astrophysicist Jocelyn Bell Burnell, who co-discovered the first radio pulsar, aptly stated, 'these radio signals are a message from the cosmos, but one that we have yet to decipher.

429. **The Bermuda Triangle of Space: A Mysterious Triangular Structure.** Similar to the Bermuda Triangle on Earth, a triangular structure has been observed in space. This puzzling formation raises many questions, such as its origin and cause, and scientists have yet to fully understand it. It may be the result of a collision between two celestial bodies, a natural formation, or an unknown phenomenon. By studying this triangular structure, scientists hope to gain valuable insights into the formation and evolution of the universe.

430. **The Kessler Syndrome: A Space Disaster Waiting to Happen**. The Kessler Syndrome is a theoretical scenario in which Earth's orbit is overpopulated with objects and debris, preventing the use of satellites in certain sections of Earth's

orbit, and increasing the chances of collisions. This hypothesis is a reminder of the importance of space debris management and the consequences of neglecting it.

"

The universe is under no obligation to make sense to you.

– Neil deGrasse Tyson

SECTION 25
Subatomic World: Quantum Mechanics and Antimatter

431. **S**uperposition: The Phenomenon of Particles Existing in Multiple States Simultaneously. Quantum mechanics allows particles to exist simultaneously in multiple states or locations simultaneously, a phenomenon known as superposition. This means that a particle can be in two or more states simultaneously until it is observed or measured. This has been experimentally verified in numerous quantum experiments and has important implications for the development of quantum technologies such as quantum computing.

432. **Q**uantum Entanglement: Instantaneous Information Transfer Across Distances. Entanglement is a phenomenon in quantum mechanics where particles can become "entangled," causing the state of one particle to instantaneously affect the state of another particle, regardless of distance. This strange and mysterious connection has been confirmed in numerous experiments and has been explored as a potential means of communication in the development of quantum communication networks. It is considered one of the most bizarre quantum phenomena, and its instantaneous information transfer has earned the moniker "spooky action at a distance" by Albert Einstein. Measuring the state of a quantum particle not only determines its properties but also breaks the entanglement. Recently, scientists have demonstrated entanglement between particles with opposite charges, leading to a new way of observing the nucleus of an atom. This opens up exciting possibilities for discoveries and applications in the field of quantum mechanics.

433. **Uncertainty Principle: Limiting Our Knowledge of Particle Properties.** The uncertainty principle is a fundamental concept in quantum mechanics that limits our knowledge of a particle's properties. It states that the more accurately we know the position of a particle, the less we can know about its momentum, and vice versa. In essence, it is impossible to know both the position and momentum of a particle with absolute certainty. This principle has been experimentally verified and has significant implications for the behavior of quantum systems.

434. **Quantum Tunneling: Particles Passing Through Barriers.** Quantum mechanics predicts that particles can tunnel through barriers that they should not be able to penetrate, according to classical physics. This phenomenon, known as quantum tunneling, has been observed in numerous experiments and has important implications for fields such as electronics and nanotechnology.

435. **Observer Effect: How Observation Alters Particle Properties.** In quantum mechanics, particles only have definite properties once they are observed or measured. This is known as the observer effect or measurement problem. The observation or measurement of a particle collapses its wave function and forces it to take on a definite state. This has been a topic of philosophical debate in the scientific community and has important implications for the interpretation of quantum mechanics.

436. **Wave-Particle Duality: Particles and Waves Behaving Alike.** According to the wave-particle duality theory, particles and waves can display each other's properties. This contradicts classical mechanics or Newtonian physics. Quantum mechanics predicts, and experimental results have verified, that particles and waves can exhibit both wave-like and particle-like behaviors. This duality is called wave-particle duality and has significant implications for the behavior of quantum systems.

437. **Quantum Teleportation: Transmitting Particle State to Another Location.** Quantum teleportation allows for the transmission of quantum information from a sender to a receiver situated some distance away. Unlike what is portrayed in sci-fi, this technique transfers solely quantum information, not physical objects. This process involves destroying the state of one particle on one end and extracting it on the other end, while classical communication between observers is necessary. Quantum teleportation is useful for sending quantum information over long distances without exposing it to adverse effects. While it is theoretically possible to teleport macroscopic objects, maintaining the necessary number of entangled states is practically impossible. Quantum teleportation is essential for quantum computers, and it may eventually enable the development of a quantum internet.

438. **The Solar Neutrino Problem: The Sun's Surprising Neutrino Production.** Neutrinos are subatomic particles produced during the process of nuclear fusion in the sun. In the 1960s, scientists found that only about a third to one-half of the predicted number of electron neutrinos arrived in detectors. This problem, known as the solar neutrino problem, was studied in the Homestake experiment, which used 100,000

gallons of dry-cleaning fluid to search for neutrinos. Only one-third of the neutrinos seemed to arrive, and researchers weren't sure where the problem was. Some scientists thought that the neutrino model was the error, but many were skeptical. This "neutrino deficit" is still not fully understood, but it is theorized to be caused by some unknown properties of neutrinos or by errors in the models used to predict the number of neutrinos produced.

439. **The Elusive Neutrinos: Abundant Yet Mysterious Particles That Challenge Scientists.** Neutrinos are tiny and light particles that are difficult to study as they only interact through the weak force and gravity. They are also the most abundant massive particles in the universe, coming from various sources through a process called decay. Neutrinos have different types and flavors, including antimatter versions, with undiscovered kinds possibly existing.

440. **The Constant Presence of Neutrinos: Trillions Pass Through Your Body Every Second.** Neutrinos are created by nuclear reactions in stars and supernovae, including our sun. They come in three different types or flavors: electron, muon, and tau. As they travel through space, neutrinos can change or "oscillate" between these flavors, a phenomenon known as neutrino oscillation. Despite their tiny mass, it's estimated that about 100 trillion neutrinos pass through every square centimeter of our bodies each second.

441. **Anti-Matter: The Mirror Image of Matter.** Anti-matter is the opposite of regular matter and is composed of particles that have the opposite charge of their matter counterparts. When

matter and anti-matter come into contact, they annihilate each other and release energy. Scientists are still working to understand the properties of anti-matter, but it has the potential to be used as a powerful energy source in the future.

442. **The Mystery of the Matter-Antimatter Asymmetry: The Universe Defied Annihilation.** The Big Bang theory suggests that matter and antimatter were created in equal amounts during the formation of the universe. However, if matter and antimatter were to meet, they would annihilate each other, leaving behind nothing but energy. This should have resulted in the complete annihilation of all matter in the universe, including ourselves. Nevertheless, we exist, and physicists believe that this is because there was one extra matter particle for every billion matter-antimatter pairs created during the big bang. Despite this asymmetry, physicists continue to study and search for an explanation for the phenomenon.

443. **Antimatter is Closer Than You Think: From Cosmic Rays to Bananas and Our Own Bodies.** Antimatter is not as distant as we may think. Cosmic rays, which are energetic particles from space, constantly bombard the Earth with small amounts of antimatter. These particles can be found in our atmosphere at a rate of anywhere from less than one to over 100 per square meter. Moreover, scientists have observed the production of antimatter above thunderstorms. Interestingly, bananas also produce antimatter, with one positron, the antimatter equivalent of an electron, being released about every 75 minutes. This is due to the presence of a small amount of potassium -40, a naturally occurring isotope of potassium found in bananas. Our bodies also contain potassium -40, which means that we also emit positrons. However, since

antimatter annihilates immediately upon contact with matter, these particles are very short-lived.

444. **T**he Mysteries of Cosmic Rays: Direct Samples of Matter from Beyond our Solar System. Cosmic rays represent a rare opportunity for scientists to directly examine extraterrestrial matter. They are particles with extremely high energy that travel through space at nearly the speed of light. The majority of cosmic rays consist of atomic nuclei that have been stripped of their electrons, with hydrogen nuclei, or protons, being the most prevalent. However, we also observe other subatomic particles, such as neutrons, electrons, and neutrinos, within cosmic rays. As cosmic rays are electrically charged, they can be affected by magnetic fields, leading to deflections in their path through space, except for the highest energy cosmic rays. Due to the magnetic fields present in the galaxy, the solar system, and the Earth, cosmic rays undergo significant deviations in their journey to Earth, making it challenging to determine their exact origin. Hence, indirect methods must be utilized to determine the source of cosmic rays.

445. **N**eutrinos Could be the Key to Unraveling the Mystery of Antimatter-Matter Asymmetry. Neutrinos are unique in that they lack an electric charge, making it difficult to distinguish between a matter particle and its antimatter partner. Scientists believe that neutrinos might be their own antiparticles, known as Majorana particles. To determine whether this is true, experiments like the Majorana Demonstrator and EXO-200 are searching for a behavior called neutrinoless double-beta decay. If neutrinos are their own antiparticles, they would annihilate each other, leaving only electrons after double decay. If Majorana neutrinos are found, it could explain why there is more matter than antimatter in the universe. According to the theory, heavy Majorana neutrinos existed right after the big bang and decayed asymmetrically, leading to the small excess of matter that allowed the universe to exist. The discovery of Majorana neutrinos could potentially solve one of the biggest mysteries in the field of physics.

446. **The Hunt for Antimatter: Searching for Clues to the Universe's Existence.** Scientists are using the Alpha Magnetic Spectrometer, a particle detector on the International Space Station, to search for antimatter remnants from the Big Bang. By identifying and analyzing cosmic particles that collide and produce positrons and antiprotons, scientists hope to find evidence of antihelium atoms, which could indicate the presence of significant amounts of antimatter in the universe. The detection of even a single antihelium nucleus would provide valuable insights into the antimatter-matter asymmetry problem and the origins of our universe.

"

The history of astronomy is a history of receding horizons.

– *Edwin Hubble*

SECTION 26
Historical Events: Pioneers, Discoveries, and Milestones

447. **R**evolutionizing Our Understanding of the Solar System: The Legacy of Nicolaus Copernicus. Claudius Ptolemy, the ancient Greek astronomer and mathematician, established a geocentric model of the solar system in which the Earth was at the center and the sun, stars, and other planets revolved around it. Despite being incorrect, his Ptolemaic system remained in place for centuries and had a significant impact on Western intellectual thought. Ptolemy authored several scientific texts, including the Almagest, which expanded upon Hipparchus' geometric model of celestial motions, utilizing epicycles and eccentric circles. He cataloged 48 constellations and presented tables of information for predicting planetary locations. Although his model was eventually replaced by the heliocentric view of the solar system, Ptolemy's writings remained authoritative for over 1,200 years.

448. **K**epler's Reformation: Discovering Elliptical Orbits and Laws. Johannes Kepler, a 16th-17th century astronomer, used the precise planetary measurements of Danish astronomer Tycho Brahe to revolutionize the Copernican view of the solar system. Kepler discovered that planets move in elliptical orbits around the sun, not in perfect circles as previously thought, and formulated three laws that are still used by astronomers today. Despite facing opposition from closed-minded individuals, Kepler's work established him as a leading figure in the Scientific Revolution and his contributions to astronomy continue to be celebrated.

449. **G**alileo Galilei: Discoverer of Jupiter's Moons and Saturn's **Rings, Inventor, Persecuted.** Galileo Galilei, an Italian astronomer, physicist, mathematician, and philosopher, is renowned for his improvements to the optical telescope. He built his first telescope in 1609, which could magnify objects three times, and later that year, he created a telescope that could magnify objects up to twenty times. Galileo observed the heavens through his telescope, discovering the four primary moons of Jupiter, now known as the Galilean moons, and the rings of Saturn. In addition to his astronomical discoveries, Galileo also made notable contributions to physics, such as proving that all falling objects fall at the same rate, regardless of mass, and inventing the first pendulum clock. Despite defending Copernicus's model of the Earth orbiting the sun, which Kepler had already calculated, Galileo faced house arrest at the end of his life due to his beliefs.

450. **I**saac Newton: The Renowned English Astronomer Who **Revolutionized the Field of Science.** Sir Isaac Newton, the renowned English astronomer (1643–1727), revolutionized the field of science through his groundbreaking work on forces, particularly gravity. By building upon the work of his predecessors, he derived three laws governing the motion of forces between objects, which are now known as Newton's laws. These laws describe how objects at rest or in motion will behave unless acted upon by an external force, how the net force acting on an object is related to its acceleration, and how every action has an equal and opposite reaction. Newton also contributed to the fields of optics, mechanics, experimental chemistry, alchemy, and theology, and is credited with inventing calculus. His groundbreaking work on universal gravity permanently altered the field of science.

451. **C**harles Messier: The French Astronomer Who Cataloged Celestial Objects and Discovered Comets. Charles Messier, a French astronomer who lived from 1730 to 1817, compiled a catalog of celestial objects, originally referred to as "nebulae," which contained 103 entries upon its final publication. Although he later added additional objects to the list based on his notes. Many of these objects are now commonly identified by their catalog names, such as the Andromeda Galaxy, known as M31. In addition to his catalog, Messier also discovered 13 comets during his lifetime. His interest in astronomy developed early, having observed a 6-tailed comet at age 14 in 1744 and an annular solar eclipse in 1748.

452. **A**lbert Einstein: Revolutionizing Our Understanding of the Universe with His Theories. Albert Einstein, a German physicist, revolutionized our understanding of the universe in the early 20th century with his groundbreaking theories. His ideas proposed that the laws of physics are universal, the speed of light is constant, and space and time are interconnected as a single entity called space-time, which is affected by gravity. These concepts fundamentally changed the field of physics, and Einstein is widely regarded as one of the greatest scientists of all time.

453. **E**dwin Hubble: The Astronomer Who Discovered the Expanding Universe. American astronomer Edwin Hubble (1899- 1953) discovered the existence of a small object in the sky that existed outside the Milky Way. Before his discovery, there was a debate regarding the size of the universe and whether it only contained one galaxy or more. However, Hubble's observation led him to conclude that the universe was expanding, which later became known as Hubble's law. He further classified the galaxies, creating a standard system of classification still in use today.

454. **Stephen Hawking: A Brilliant Mind and Communicator Who Transformed Our Understanding of the Universe.** Stephen Hawking, a renowned cosmologist and theoretical physicist, made significant contributions to the field of cosmology. He proposed that the universe has a beginning and an end, with no boundary or border. Despite being considered one of the greatest scientific minds since Einstein, Hawking directed his books and lectures toward the general public, educating them about the universe. Despite being afflicted with motor neurone disease since he was 20, Hawking completed his doctorate in cosmology at Cambridge. His primary discovery was that the universe began at the Big Bang and will come to an end. Hawking also predicted that black holes emit radiation known as Hawking radiation. He chronicled his discoveries in several books, including the best-selling "A Brief History of Time."

455. **Carl Sagan: The Charismatic Astronomer Who Popularized Space Science.** Carl Sagan, an American astronomer born in Brooklyn, New York in 1934, was renowned for his influential work in popularizing astronomy. Though not considered the most accomplished scientist in comparison to some of his contemporaries, Sagan made significant contributions to planetary science and broke down complicated subjects in a way that interested television viewers while educating them. His boundless energy and charismatic teaching style inspired people globally, and he founded the Planetary Society, a nonprofit organization aimed at advancing space science and

exploration. Sagan also made notable scientific discoveries, such as explaining the high temperatures of Venus and the seasonal changes on Mars, while serving as a professor of astronomy and space sciences and director of the Laboratory for Planetary Studies at Cornell University.

456. **The First Liquid-Fueled Rocket Launch in History: Robert H. Goddard's Groundbreaking Achievement.** On March 16, 1926, American Robert H. Goddard launched the world's first liquid-fueled rocket, which lasted for 2.5 seconds and reached an altitude of 41 feet, traveling at a speed of 60 mph. This historic rocket was 10 feet tall, made of thin pipes, and fueled by gasoline and liquid oxygen.

457. **The First Artificial Satellite in History and the Start of the Space Age: Sputnik-1.** On October 4, 1957, the Soviet Union launched Sputnik-1. Weighing 183.9 pounds and measuring only 22.8 inches in diameter, the satellite orbited the Earth every 98 minutes. This event marked the beginning of the space age and ignited the space race between the United States and the Soviet Union, ultimately leading to the establishment of NASA. At the Library of Congress and online, there are resources available for further research on this topic.

458. **The First Human in Space: Yuri Gagarin's Groundbreaking Journey.** On April 12, 1961, Russian cosmonaut Yuri Gagarin became the first human in space, orbiting Earth for 108 minutes in his Vostok 1 spacecraft. Gagarin achieved cultural hero status in the Soviet Union after his historic flight on the Vostok spacecraft. Over six decades later, he continues to be widely commemorated in Russian space museums, where

a plethora of artifacts, busts, and statues pay tribute to him. Gagarin's burial place is located at the Kremlin in Moscow, and a portion of his spacecraft is showcased at the RKK Energiya Museum.

459. **Alexei Leonov's Historic Spacewalk: First Steps in Outer Space.** In 1965, Alexei Leonov became the first human to walk in space, a historic moment in space exploration. As he exited his spacecraft, Leonov famously exclaimed, "The Earth is round!" The experience was marked by an extraordinary silence, as Leonov floated in the vastness of space.

460. **Apollo 11 and The First Humans on the Moon: Edwin "Buzz" Aldrin and Neil Armstrong.** In 1969, the historic Apollo 11 mission saw American astronauts Edwin "Buzz" Aldrin and Neil Armstrong become the first humans to set foot on the moon. Millions of people around the world tuned in to watch as the two astronauts, sporting bulky space suits and backpacks of oxygen to breathe, made their way onto the moon's surface. Approximately six-and-a-half hours after landing, Armstrong took his first steps on the lunar surface. The Apollo 11 mission was the culmination of a national goal set by President John F. Kennedy eight years earlier, announcing the ambitious mission of landing a man on the moon by the end of the 1960s. The final manned moon mission, Apollo 17, took place in 1972.

461. **Hubble Space Telescope: Capturing the Universe's Earliest Galaxies from 13.4 billion light-years away.** The Hubble Space Telescope, launched by the space shuttle Discovery on April 24, 1990, is a large telescope located in space that orbits approximately 535 kilometers (332 miles) above

Earth. As long as a large school bus and weighing as much as two adult elephants, it is powered by solar energy and travels at a speed of approximately 5 miles per second. The telescope's main function is to capture clear images of planets, stars, and galaxies, and it has recorded over one million observations to date. Unlike ground-level telescopes, Hubble bypasses atmospheric distortion and obstruction, providing a better view of the universe. It can capture images from up to 13.4 billion light-years away, allowing scientists to see some of the earliest galaxies in the universe. Through these observations, scientists have been able to learn about the birth and death of stars and galaxies billions of light-years away, as well as comets crashing into Jupiter's atmosphere.

462. **Kepler Space Telescope's Legacy: Discovering 2,662 Earth-Sized Exoplanets.** The Kepler Space Telescope, named after astronomer Johannes Kepler, was launched by NASA in 2009 to discover Earth-sized planets that orbit other stars. The telescope had a single scientific instrument, a photometer, which monitored the brightness of around 150,000 stars in a fixed field of view. Kepler's mission was to search for exoplanets in the Milky Way galaxy, specifically ones similar in size to Earth and located in or near habitable zones. Over the course of almost a decade, Kepler observed over 530,000 stars and detected a total of 2,662 exoplanets by analyzing data that was transmitted back to Earth. The Kepler Space Telescope was retired in 2018 after running out of fuel.

463. **The Legacy of NASA's Cassini Spacecraft: Exploring the Wonders of Saturn.** Cassini was a spacecraft that explored Saturn and its icy moons for over a decade, revealing detailed

information about the planet's storms and gravity. It also carried the Huygens probe, the first human-made object to land on a world in the distant outer solar system. After 20 years in space, including 13 years exploring Saturn, Cassini was sent on a final mission to protect moons that could have conditions suitable for life. It made nearly two dozen dives before plunging into Saturn's atmosphere in 2017, returning scientific data until its very end. Cassini's mission was significant in advancing our understanding of Saturn and its potential for life.

464. **Voyager 1 and 2: The Most Distant Human-Made Object in Space, Now Over 14 Billion Miles Away.** Voyager 1, launched more than 20 years ago, is the most distant human-made object in space, having cruised beyond the Pioneer 10 spacecraft, at 10.4 billion kilometers from the Sun. Both Voyager 1 and its twin Voyager 2 were launched in 1977 and have flown by Jupiter, Saturn, Uranus, and Neptune. They are currently journeying into the space between the stars, with Voyager 1 becoming the first spacecraft to leave the solar system in 2012. As of April 2023, Voyager 1 is over 14 billion miles away from Earth, and it sent back a message in 2021 that it detected a faint hum of interstellar space. Interestingly, for a few months every year, the Voyager spacecraft gets closer to Earth because of our orbit around the sun.

465. **Curiosity Rover: The Unstoppable Martian Explorer-Breaking Records with a Decade-Long Mission on Mars.** The Curiosity Rover, the most advanced and sizable rover ever launched to Mars, experienced a successful landing on August 5, 2012, after enduring the perilous "seven minutes of terror" landing process. Initially intended to operate for a

mere two years, the rover's exceptional accomplishments led to its mission being extended indefinitely, rover has traveled over 25 kilometers (15.5 miles) on the Martian surface. With its advanced scientific instruments and equipment, the rover can thoroughly examine the Martian environment and geology. Its objectives involve examining the Martian climate and geology, seeking clues of microbial life, and planning for future human expeditions. As of April 12, 2023, the rover had been functioning on Mars for a decade and 249 days.

466. **S**pirit and Opportunity: The Rovers That Revealed an Ancient Watery Past on Mars. In 2004, two robots named Spirit and Opportunity were sent to Mars to explore the Martian surface, study the planet's atmosphere, and search for evidence of past water activity. The rovers successfully landed on opposite sides of the planet and conducted field geology, sending high-resolution images and data back to Earth. The mission's primary objective was to find rocks and soils that could provide evidence of past water activity on Mars. The rovers found evidence of ancient Martian environments where intermittently wet and habitable conditions existed. Spirit and Opportunity exceeded their planned 90-day mission lifetimes by many years, sending thousands of images and discovering a variety of rocks indicating that early Mars was characterized by impacts, explosive volcanism, and subsurface water. Opportunity continues to operate, having covered a record-breaking distance of more than 26 miles (42 kilometers) on Mars.

467. **T**he Golden Record: A Message from Earth to the Universe. In 1977, NASA launched Voyager space probes 1 and 2 to explore the outer solar system, and the Golden Record was

included on board each spacecraft. The Golden Record serves as a message to any extraterrestrial life that may encounter the probes. It contains images and sounds that depict the diversity of life and culture on Earth and is considered a time capsule. Despite being over 11 billion miles from Earth as of 2020, it will take another 40,000 years for either probe to encounter another planetary system. Both probes have reached interstellar space, and Voyager 1 is currently the farthest human-made object from Earth. The Golden Record, a 12-inch gold-plated copper disk, is designed to tell the story of humanity's world to any extraterrestrial beings who may come across it.

468. **The Floating Lab in the Sky: The International Space Station.** The International Space Station (ISS) is a collaborative project between multiple countries and has been continuously occupied since 2000. It is the largest human-made structure in space, roughly the size of a football field, and has been continuously occupied by astronauts since 2000. It serves as a platform for a variety of scientific experiments and research in fields such as physics, biology, and astronomy. The station orbits the Earth at an altitude of about 408 km and completes one orbit in about 90 minutes. The International Space Station, which travels at nearly 17,398 miles per hour (28,000 kilometers per hour), is perceived by us as a swiftly moving star, but its visibility depends on both our location and the station's orbit, which dictates when and where we should search for it.

469. **The Quest for Extraterrestrial Life: The SETI Institute.** The SETI (Search for Extraterrestrial Intelligence) Institute is an organization dedicated to the search for intelligent life in the universe. The institute uses various methods, such as radio

telescopes and infrared telescopes, to scan the cosmos for signs of extraterrestrial life. The institute has been involved in many exciting projects, including the development of a new tool called Laser SETI, which uses lasers to search for flashes of light in the sky as potential signals of intelligent life.

"

"The Earth is the cradle of humanity, but mankind cannot stay in the cradle forever.

– Konstantin Tsiolkovsky

SECTION 27
Exploring the Future Frontier: Space Tourism and Beyond

470. **Space Tourism: A Growing Industry.** Several private companies founded by Jeff Bezos (of Amazon), Richard Branson (Virgin Group), and Elon Musk (of Tesla and SpaceX) are transforming the space tourism industry by developing space tourism programs, to eventually make space travel accessible to a broader audience. Jeff Bezos' Blue Origin is creating a reusable rocket system called New Shepard to make space travel accessible to 'everyone'. Richard Branson's Virgin Galactic is developing a spacecraft called VSS Unity that could take passengers on suborbital flights. Elon Musk's SpaceX has developed the Falcon 9, a reusable rocket system that has already sent astronauts to the International Space Station and is planning to launch its first space tourists later this year. While the cost of space tourism is currently prohibitively expensive for most people, it is possible that advances in technology and economies of scale could eventually make it more affordable.

471. **Artemis Program: NASA's Moon to Mars Mission.** The Artemis program is NASA's ambitious plan to land humans on the Moon by 2024 and establish a sustainable presence on the lunar surface by 2028, using it as a stepping stone for future manned missions to Mars and other deep-space destinations. The program aims to "land the first woman and first person of color on the Moon," explore the lunar surface, and lay the groundwork for sending astronauts to Mars. Artemis II, scheduled for November 2024, will have a four-person crew orbiting the Moon but not landing on it.

472. **SpaceX: The Company Revolutionizing Space Travel and Pursuing the Colonization of Mars.** SpaceX is an American company founded in 2002 by Elon Musk that specializes in spacecraft manufacturing, launching, and satellite communications. Its primary mission is to reduce the cost of space transportation and enable the colonization of Mars. The company has developed a range of launch vehicles including Falcon 9, Falcon Heavy, and Starship, along with rocket engines, spacecraft, and communication satellites. SpaceX's major projects include Starlink, a satellite internet constellation that has over 3,300 small satellites in orbit, and Starship, a fully reusable super heavy-lift launch system with the highest payload capacity of any orbital rocket. Despite experiencing setbacks like Starship's first launch "failure", SpaceX has made notable achievements in space exploration, including being the first private company to launch, orbit, and recover a spacecraft, sending a spacecraft to the International Space Station, achieving vertical propulsive landing of an orbital rocket booster, reusing such a booster, and sending astronauts to orbit and the International Space Station. With over one hundred launches and landings of Falcon 9 rockets, SpaceX is at the forefront of the race to colonize Mars.

473. **Starship: The Most Powerful Rocket Ever Built to Transport Humans to the Moon and Mars.** Starship is the successor to SpaceX's previous rockets, including Falcon 1, Falcon 9, and Falcon Heavy, and the world's most powerful rocket ever built, with the Super Heavy launch vehicle capable of lifting 100,000 kg (220,000 pounds) for long-duration interplanetary flights. With a height of 400 feet (122 meters), Starship is the tallest rocket ever built, surpassing the Saturn V, which stood at 363 feet (110.643 meters) tall. The system is fully reusable, making it the first of its kind. Starship comes in two configurations: one for carrying cargo, which can accommodate payloads up to 22 meters high, and one for carrying up to 100 astronauts.

474. **N**ASA's **Mission to Prepare for Human Exploration: Four Humans Set to Spend Year on Simulated Mars Habitat.** NASA will be sending four humans to Mars in June 2023 for a year-long mission to prepare for future human exploration of the Red Planet. The crew will live in a simulated habitat that includes private quarters, a kitchen, and areas for recreation, fitness, work, and crop growth activities. They will carry out simulated spacewalks, robotic operations, habitat maintenance, personal hygiene, exercise, and crop growth activities while facing environmental stressors such as resource limitations, isolation, and equipment failure. The analog mission is the first of three planned by NASA to better understand the requirements for a habitat on Mars.

475. **C**olonizing **Mars: SpaceX's Bold Plan to Create a Self-Sustaining Human Colony on the Red Planet.** The great visionary Elon Musk (from Tesla and SpaceX) plans to establish a self-sustaining human colony on Mars within just 20 years. His company, SpaceX, has been relentlessly pushing the boundaries of space exploration, with the ultimate goal of making humanity an interplanetary species. The prospect of colonizing Mars is not just a scientific achievement, but also a historic moment that would change the course of human history forever. It's a bold, daring, and thrilling adventure that has captured the imagination of millions around the world.

476. **B**reathing on Mars: NASA's Perseverance Rover Extracts Oxygen from the Red Planet's Atmosphere. NASA's Perseverance rover has successfully demonstrated the possibility of generating oxygen from Mars' atmosphere through an experimental device called the Mars Oxygen In-Situ Resource Utilization Experiment (MOXIE). Using a process called electrolysis, MOXIE breaks down the carbon dioxide in the Martian atmosphere into its component elements of oxygen and carbon monoxide. This breakthrough technology could provide a vital resource for future human exploration efforts on Mars, allowing astronauts to breathe and even create rocket fuel using the oxygen generated by MOXIE.

477. **T**he Most Valuable Asteroid in Space: NASA Discovers a $10,000 Quadrillion Metallic Wonder. One of the most fascinating discoveries in space exploration is the metallic asteroid named '16 Psyche'. NASA's Hubble Space Telescope has recently found this asteroid located in the main asteroid belt between Mars and Jupiter, approximately 370 million kilometers away from Earth. With a diameter of 226 kilometers, it is roughly the size of West Virginia. The asteroid is said to be worth a staggering $10,000 quadrillion, which is more than ten thousand times the global economy in 2019. Although the exact composition of Psyche is still unclear, researchers believe that its surface could be made of pure iron with a high concentration of nickel. This discovery has opened up new avenues for research, and scientists are excited to explore the unique properties of this metal-rich asteroid.

478. **T**he Race to Mine Space: Advancements in Technology and Private Sector Competition Ignite a New Frontier in Resource Exploration. The advancements in technology, falling costs, and private sector competition have put space exploration back at the forefront, with commercial space mining on the horizon. NASA has already awarded contracts to four companies to extract lunar regolith by 2024, marking the start of commercial space mining. The economic viability of the industry depends on major technological, financial, and business model advancements, as well as the governance of space resources. The potential benefits of space mining have sparked geopolitical competition, with several countries pursuing space mining ambitions. The Moon is a prime target for commercial mining due to its relative closeness, low gravity, and ability for remote operation.

479. **Asteroid Mining: A Revolutionary Solution for Resource Acquisition and the Future of Space Economy.** Asteroid mining has the potential to revolutionize resource acquisition and shape the future of the space economy. As environmental degradation and resource depletion become more prevalent, asteroid mining could provide a way to obtain valuable resources while eliminating the need for in-the-ground mining methods and reducing inhumane and illegal mining practices. However, the economic impact of asteroid mining is a subject of debate. Some experts believe it could create significant wealth and lead to the development of a space economy, while others argue that it could harm countries that rely on resource exports, such as South Africa, and even destroy the global raw material economy. It's important to note that asteroid mining is still a relatively new field, and there is much to be learned about it.

480. **Revolutions in Orbit: The Frequency of Satellite Passes.** The speed at which a satellite orbits the Earth depends on

the height of its orbit and the size of its path. Low Earth orbit satellites, such as the International Space Station, complete a full rotation around the Earth in about 90 minutes. On the other hand, geostationary satellites, utilized for communication and weather prediction, circle the Earth once every 24 hours.

481. **The Growing Problem of Human-Made Debris in Orbit: The Growing Threat of Space Junk.** Space pollution, also called "space junk," is a rising danger to space exploration and satellite technology. With an increase in satellites and space missions, the risk associated with space debris becomes increasingly urgent. The growing number of space debris and satellite mega-constellations poses a threat to the safety of astronauts, space missions, and our planet. Effective space debris mitigation and responsible satellite deployment have become more necessary than ever. There are currently over 27,000 tracked pieces of debris in space, with many more that are too small to track but still dangerous. Nonfunctional spacecraft, abandoned launch vehicle stages, mission-related debris, and fragmentation debris all contribute to the rising population of space debris. Scientists use radar to monitor objects larger than 10 centimeters, but there are tens of millions of pieces smaller than 1 cm. An estimated 31,870 debris objects larger than a softball will orbit Earth in 2023, with a staggering 131 million untracked pieces thought to exist. Even a small collision with debris can cause significant damage to spacecraft and human spaceflight, making the threat of space debris very serious.

482. **Space Tourism's Impact on Climate: Study Warns of Significant Threat to Ozone Layer and Global Warming.** According to a new study published on April 13, 2022, in the journal Geophysical Research Letters, space tourism may have a significant impact on global warming and the Earth's protective ozone layer. The study warns that rockets launched by billionaires like Richard Branson and Elon Musk release black carbon into the stratosphere, which is 500 times more harmful to the climate than on Earth. While Jeff Bezos' rockets burn liquid hydrogen and oxygen, which is less damaging, all rockets contribute to ozone depletion. The study suggests that space tourism is projected to account for 6% of warming due to black carbon emissions, even though it only contributes 0.02% of global emissions, raising concerns about the billionaire space race and the need for the regulation of space tourism before it becomes an untenable problem for our planet.

"

Space is big. You just won't believe how vastly, hugely, mind-bogglingly big it is. I mean, you may think it's a long way down the road to the drug store, but that's just peanuts to space.

– Douglas Adam

SECTION 28
Space Oddities: Eccentric and Curious Journey

483. **T**ears in Space: Zero Gravity Affects Emotional Expression. In the vacuum of space, tears behave differently due to the absence of gravity. Tears will still form in your eyes, but without gravity, they will not fall on your face. Instead, they will cling to your eye and form a bubble. This means that tears in space will not provide the same emotional release as on Earth. The bubble will stay in place until it becomes too big to hold or is removed by wiping it away. So, in space, you can still cry, but your tears won't have the same dramatic effect as they do on Earth.

484. **B**eauty in the Absence of Gravity: The Artistic Symmetry of Free Liquids in Space. In the absence of gravity, liquids behave in a unique way known as surface tension, causing them to form perfectly spherical shapes. This phenomenon is visible in the exquisite water droplets aboard the International Space Station, where the absence of gravitational forces allows the liquid to take on a beautifully symmetrical and spherical form. This breathtaking display of the artistic side of physics and the universe is akin to wandering through a cosmic art exhibit, captivated by the mesmerizing symmetry of it all.

485. **A**stronauts' Spine Stretching: Spinal Elongation in Space. Astronauts can experience a slight increase in height in space due to the absence of gravity, called "spinal elongation." The lack of gravity allows the intervertebral discs to expand and cause the spine to stretch out, leading to an increase in height, typically around 2 inches or 5 cm. However, this effect is temporary, and the height gain is not permanent; once the astronaut returns to Earth, gravity reasserts its impact on the spine, and the astronaut will return to their average height. Nevertheless, not every astronaut experiences this effect.

486. **L**EGO **Minifigures Sent to Explore Jupiter: Still Orbiting the Planet Today.** Three LEGO minifigures were sent to orbit Jupiter aboard the Juno space probe, to spark children's interest in science and space exploration. The minifigures depict Galileo Galilei, Jupiter, and Juno, and were made from spacecraft-grade aluminum to withstand the harsh conditions of space. These figures have no movable parts and are still orbiting the planet as part of the Juno mission, which has been extended until September 2025.

487. **L**ost and Found in Space: The Story of a Wedding Ring during the Apollo 16 Mission.** During the Apollo 16 lunar mission, command module pilot Ken Mattingly lost his wedding ring, which was eventually found on the ninth day of the mission. The ring was spotted floating out of the hatch door during a spacewalk and was caught by another astronaut, Charles Duke Jr. This incident highlights the fact that even small objects can be easily lost in space, and there are already many items floating in the vast cosmic expanse, including a toothbrush.

488. **S**pace Gardening: Growing Plants in Space.** NASA has conducted several experiments on growing plants in space to understand how plants respond to microgravity and radiation. The goal is to eventually develop space gardening systems that can provide food for astronauts during long-duration missions. Plants have been grown on the ISS, including lettuce, radish, and zinnias. In 2021, the first-ever space tomatoes were harvested on the ISS, demonstrating the possibility of growing crops in space.

489. **M**artian Blueberries: Abundant Iron-Rich Pebbles with a Watery Past. Hematite spherules, also known as the Martian 'blueberries', are small pebbles that contain a high amount of iron oxide and are abundant in Meridiani Planum on Mars. These spherules come in varying sizes and can either be embedded in sediments or loose on the surface. They were discovered by NASA's Mars Exploration Rover Opportunity in 2004 and are called "blueberries" due to their bluish-grey color. The formation of these spherules involved the flow of acidic, salty, liquid water over two geological epochs. Determining the amount of iron oxide present in the blueberries has been challenging.

490. **A** Man Buried on the Moon: Eugene Shoemaker. Eugene M. Shoemaker, a renowned planetary geologist who expressed his unfulfilled dream of visiting the moon, has been honored with a unique tribute. He is the only person to have his remains buried on the moon, as his ashes were placed inside a polycarbonate capsule and launched aboard the Lunar Prospector spacecraft. The small capsule, measuring only one-and-three-quarters inches in length and seven-tenths of an inch in diameter, is securely stored in a vacuum-sealed aluminum sleeve within the spacecraft. This tribute reflects Shoemaker's immense contribution to planetary science and serves as a symbol of humanity's fascination with space exploration.

491. **The Groundbreaking Voyage of Laika: The First Mammal in Space.** In November 1957, the Soviet Union made history by sending a dog named Laika into orbit on the Sputnik 2 satellite. Although Laika was the first mammal to be sent into space, she was not the first animal. The United States had previously launched fruit flies on a suborbital mission in February 1947. Laika was chosen from the many stray dogs rescued from the streets to participate in the Soviet spaceflight program. The program only selected female dogs, believing them to be better suited for confinement. Laika underwent extensive training to prepare her for life in space, including adapting to smaller living spaces and eating jellied food to accommodate weightlessness. The mission was known to be fatal for Laika, but the true details of her fate were not revealed until years later.

492. **Valeri Polyakov: The Cosmonaut Who Paved the Way for Long-Term Spaceflight.** On March 22, 1995, Valeri Polyakov, a Russian cosmonaut, established a new record for the longest single spaceflight in human history. Polyakov returned to Earth after a 437-day space mission on the Mir space station. Astonishingly, he did not experience any significant physical or mental impairments upon returning. Researchers took note of these findings and concluded that astronauts could maintain stable moods and overall function during prolonged spaceflights, such as manned missions to Mars.

493. **Arthur C. Clarke's Collaboration with Kubrick: Pioneering the Science Fiction Genre 2001: A Space Odyssey.** Stanley Kubrick's 1968 masterpiece was a revolutionary film that was ahead of its time and presented many scientific concepts. His cinematic depiction of a rotating space station simulating artificial gravity and his HAL-9000 exploration of artificial intelligence were groundbreaking and made a significant contribution to the science fiction genre and the film industry. Kubrick's collaboration with noted futurist and author Arthur C. Clarke helped ensure the film's academic accuracy. Clarke, who wrote the short story that inspired the film The Sentinel, had expertise in space travel and technology that allowed him to ground the scientific concepts of cinema into reality. The film explored humanity's relationship with technology and the unknown, creating a timeless classic that continues to influence and inspire to this day.

494. **NASA and Star Wars: Strong Connections Between Science Fiction and Real-World Space Technology.** NASA and Star Wars share real-world technologies, such as ion engines, and have observed planets similar to those found in the science fiction franchise. The latest Star Wars film was sent to the International Space Station for astronauts to watch. Astronaut Kjell Lindgren even posed in a Jedi-themed mission poster with his crewmates and pointed out the similarity of the station's cupola to the cockpit of an Imperial TIE Fighter. NASA is already using ion engines on spacecraft and considering them for future missions, including the Asteroid Redirect Mission. Interested in space exploration? Apply to be an astronaut with NASA and become a "sky walker."

495. **Space Beer: The First Beer Brewed With Moon Dust.** In 2019, a group of researchers brewed the first beer made with moon dust using simulated lunar soil. The beer, called "Lunar Oktoberfest," was brewed in honor of the 50th anniversary of the Apollo 11 moon landing and was described as having

a "smoky and spicy flavor." While not currently available for consumption, the experiment shows the potential for using local resources for food and beverage production on future lunar missions.

496. **W**ow! **Signal: The Mysterious Message from the Depths of the Universe.** In 1977, the Big Ear Radio Telescope in Ohio picked up an incredibly strong signal that lasted only 72 seconds. An astronomer who noticed the signal days later marked the page with a red "Wow!" due to its resemblance to what SETI researchers expected to see from an alien intelligence. Despite numerous attempts to track the signal, it has never reappeared. This event, known as the "Wow! Signal," has become a widely discussed topic in the search for extraterrestrial life, with some seeing it as the most promising indication of alien life while others view it as more of a publicity stunt than a scientific discovery.

497. **B**uzz **Lightyear's Epic Odyssey: The Toy Astronaut From a Disney Movie Goes on a Long Space Journey.** Buzz Lightyear, the fictional toy character from the Disney movie "Toy Story," has flown in space for over 400 days and holds the record for the longest time spent on a single space mission at 468 days. Despite being a toy, Lightyear has become an icon, loved by both children and adults, and has even been selected as a space education advocate by NASA and Disney Parks. The recent release of the movie "Lightyear" has once again brought the beloved character to the forefront of popular culture.

498. **T**he **Lasting Legacy of Apollo: Footprints on the Moon Will Remain Preserved for Millions of Years.** The lack of wind and atmosphere on the Moon ensures that the footprints, such as

Neil Armstrong's first step on the lunar surface, remain as crisp as the day they were made, a lasting testament to humanity's first steps on another celestial body. Their longevity may even outlast our own civilization. These iconic footprints serve as the ultimate "out of this world" souvenir.

499. **The Pale Blue Dot: How a Single Photograph Changed Our View of Earth's Place in the Universe.** The term "pale blue dot" refers to the photograph captured by the Voyager 1 spacecraft in 1990, which depicts our home planet as a tiny, blue speck in the vast expanse of space. Taken from an unimaginable distance of 3.7 billion miles (6 billion kilometers), the image shows Earth as a minuscule dot of light surrounded by the emptiness of space. Astronomer Carl Sagan coined the term to describe the photo's unique perspective on our planet's place in the cosmos. In his book, also titled "The Pale Blue Dot," Sagan explores the significance of this image and its impact on humanity's understanding of its role in the universe. This phrase has since become a powerful metaphor, highlighting the preciousness and vulnerability of life on Earth and the importance of protecting our planet for future generations.

500. **The Iconic First Words Spoken on the Moon: Neil Armstrong's Famous Phrase.** Neil Armstrong placed his left foot on the moon's surface at 10:56 p.m. ET on July 20, 1969, and uttered the now-famous words, "That's one small step for man, one giant leap for mankind." Since then, many people have been curious about the inspiration behind his choice of words for this historic moment.

REFERENCES

https://www.worldometers.info

https://www.space.com

https://www.science.nasa.gov

https://www.voyager.jpl.nasa.gov

https://www.solarsystem.nasa.gov

https://www.mars.nasa.gov

https://www.nasa.gov

https://www.spacex.com

https://www.jwst.nasa.gov

https://exoplanets.nasa.gov

https://www.science.org

Made in the USA
Columbia, SC
10 October 2023

24243580R00167